PONDERING RELATIVITY

An Illustrated Guide for Building an Understanding of Einstein's Relativity

Printed in the United States of America
Funke, Douglas
Pondering Relativity: An Illustrated Guide for Building an Understanding of Einstein's Relativity
ISBN: 978-0-9988811-3-3
1. Pondering Relativity 2. Douglas Funke 3. Einstein's Relativity 4. Special Relativity
5. General Relativity 6. Science and Religion 7. No Frills

All illustrations are by the author with graphic support from Seth Triggs.
Book cover by Seth Triggs and Anne Johnston Fera.

Book layout and design by Anne Johnston Fera.
Editorial support provided by Laura Balcom
The Constant Writer
Youngstown, NY.

NFB
<<<<>>>>
No Frills Buffalo/Amelia Press
119 Dorchester Road, Buffalo, NY 14213
For more information visit nofrillsbuffalo.com

This book is dedicated to my wife, Dolores,
who has been my loving companion and
best friend for over 35 years, and our three
kids, Carolyn, Christopher, and Adam.
They have been a great source of pride
and learning for us.

Acknowledgements

I want to thank my Mother, who took me on walks as a small boy to look at the stars. There were very few city lights at the time where we lived, so the sky was spectacularly lit with thousands of stars and an occasional shooting star. This ignited my interest in astronomy even before I started kindergarten. I also want to thank my Dad, who who bought me a children's telescope that I used to look at the Moon and planets, and made up for a lack of astronomy classes in grade school by buying me children's books on astronomy that I read cover to cover. This initial interest in astronomy led me to read many books on the subject throughout my adult life and was, ultimately, one important motivation for writing this book.

I want to thank my sister, Barb Keough. She has been my friend and a source of support throughout my life. She believed in me even during my younger years when I lacked the apparent elements one would need for success — direction, motivation and confidence. Most recently, she provided encouragement for me to write this book and spent many hours reviewing early drafts, drawing on her graphic artist / human engineering background and offering suggestions for making it more understandable for non-scientists. Also, drawing on her background as an operations manager at a local advertising agency, she lined up a team of editing and book design professionals, especially Laura Balcom and Anne Johnston Fera. Thanks Barbie. And thanks Laura and Anne! So many books on relativity lose readers partway through because of hard to understand scientific language and concepts. Your professional contributions helped make this book one I hope people will be able to read; and will read cover to cover.

I also want to thank Bob Wack, my good friend and a retired systems engineering physicist. He acted as a sounding board for the ideas in this book and provided a thorough review of several drafts, making sure I didn't misrepresent any of the concepts from a physics perspective. He also suggested interesting additions and expansions to make the book better. Thanks Bob!

In addition, I want to thank my astronomy professor at Buffalo State College, Dr. John Mack, who reignited my excitement in astronomy and physics even though it was too late to change my major. Also, many thanks to all my good friends at the Unitarian Universalist Church of Amherst, NY — especially the retired men's lunch group where we sometimes discuss the latest physics discoveries and their ramifications. Thanks also to Seth Triggs, my good friend and graphic artist, for designing the cover artwork and creating several of the graphic images used in the book.

I also want to thank my wife, Dolores (again). She supported me during the long period of writing this book — which took over six years. She was supportive and provided encouragement, even when progress seemed unbearably slow. She reviewed early drafts, providing guidance and suggestions for making the book more understandable. She also drew on her background as a former high school and adjunct university math teacher to help with the math in Appendix B. Thanks Dolores!

Preface

How are space and time related? Can relativity's affect on time change the order in which events occur? What does $E=mc^2$ have to do with relativity? Isn't there anything that can go faster than light? How does relativity affect how fast we grow old? What does relativity say about what we can and cannot know? Does space really compress at speeds approaching the speed of light? These are just some of the questions of interest for understanding our Universe that are answered by the Special and General Theories of Relativity. They are made understandable in this book.

The concepts of relativity are easier to understand than conventional thinking would have us believe. This book proves the point. It uses simple coordinated graphics – sequenced throughout the book – to build an understanding in a gradual stepwise fashion. The self-reinforcing graphics help visualize the principles of relativity, building from simple observations to Einstein's elegant and revolutionary conclusions. The book applies human learning principles to make an otherwise difficult subject matter understandable and enjoyable. Mental exercises are offered in Appendix C, as tools for experiencing, internalizing and remembering the relativity principles presented. After both Special and General Theories of Relativity are laid bare, implications for how the Universe works are explored and explained. Readers will gain a new and deeper appreciation of relativity and what it says about how the Universe is made. This book delivers the experience and excitement of relativity.

This book was motivated by my lifelong interest in astronomy, physics and science, together with my love of writing and passion for making things understandable. I have over 30 years experience working as a human factors engineer at several research engineering companies. Throughout this time a major part of my job was to communicate technical information clearly. I have written scores of technical reports and articles on topics ranging from automated highways to the application of human learning principles within complex training systems. I have published articles in numerous technical magazines, conference proceedings and professional journals. As a Certified Professional Ergonomist (CPE), I also bring recognized professional expertise to this book.

Since the early 1900s when Einstein stunned the scientific world with his theories, many books have been written about relativity. But relativity remains largely a mysterious enigma understood by few outside of the technical and scientific fields. I believe one reason for this is that most books on the subject have been written using traditional approaches that don't work well for complex, non-intuitive topics like relativity. They lack clear, coordinated illustrations that work together to help the reader visualize and remember the concepts being presented. Instead, they rely mostly on text to describe the ideas, while the illustrations presented are not coordinated or integrated for incremental understanding. This makes it difficult for readers to connect and internalize the principles involved.

By contrast, this book integrates coordinated illustrations throughout. These help the reader visualize how relativity works and reinforces the concepts as the reader proceeds. Like modern approaches for teaching foreign languages that pair images with the associated language to cement understanding, this book uses pictures to convey the concepts of relativity. The result is a book that works; one that builds an enduring understanding in a way that is enjoyable and rewarding. It's enjoyable because you can see how relativity works; rewarding because you experience a sense of mastery as concepts are gradually understood.

I hope this book works for you. I hope you enjoy reading this book.

Gary

Table of Contents

Acknowledgements IX

Preface XI

Introduction 1

1.1 Overview. 2

1.2 Target audience. 3

1.3 Pondering framework. 4

1.4 What you will learn. 5

What Is Light? 13

2.1 The nature of light. 14

2.2 The size of the Universe. 16

2.3 Unexpected observations. 18

2.4 The color of light. 19

2.5 The electromagnetic radiation spectrum. 21

There Is No Up or Down in Space or Anywhere Else 23

3.1 The aether. 24

3.2 The nature of motion. 26

3.3 Incompatibility between relativity and the aether. 31

3.4 The velocity of motion is relative too. 32

The Special Theory of Relativity 35

4.1 How Special Relativity Works. 36

4.2 Why Special Relativity is called "special." 41

4.3 Observer perspective is the key. 41

Clocks....45

5.1 Clocks. 46

5.2 Light-clocks. 47

5.3 Clocks and time. 50

Time and Space....53

6.1 Time slowing. 54

6.2 Space compression. 59

6.3 Bi-directional application. 61

Shortcomings of Special Relativity and the Equivalence Principle....63

7.1 Shortcomings of Special Relativity. 64

7.2 The nature of acceleration and gravity. 66

7.3 The Equivalence Principle. 69

General Relativity: Reality in Four Dimensions....75

8.1 Doubling down boldly. 78

8.2 General Relativity requires a new kind of geometry. 79

8.3 Space and time do not exist separately. 82

8.4 Space-time becomes distorted under conditions of gravitation. 85

8.5 Acceleration and gravity are equivalent, but different. 88

8.6 Motion and space-time distortions cause changes in light's appearance. 101

8.7 Extreme gravitational conditions create extreme time slowing and space compression. 107

What Relativity Says About the Universe: Part 1 — Time, Space and Aging 109

9.1 Time and aging. 110

9.2 The universal speed limit. 115

9.3 Simultaneous events and causality. 121

9.4 The relationship between event ordering, the slowing of time and the compression of space. 125

What Relativity Says About the Universe: Part 2 — How the Universe is Made 129

10.1 Moving through space-time. 131

10.2 Relativity limits information access. 135

10.3 Time it takes for light to travel across space. 138

10.4 Exceeding the universal speed limit. 140

10.5 The appearance of compressed space or actual compressed space? 148

10.6 How objects bend space-time. 151

10.7 The structure of space-time. 153

10.8 $E=mc^2$. 155

10.9 Limits of relativity's scope and competence. 158

Where's the Proof? 161

11.1 The questioning scientific community. 162

11.2 Verifications of Einstein's predictions. 165

11.3 Modern experiments that have measured the effects of relativity. 168

11.4 Putting relativity in practice. 173

Science and Religion 175

12.1 The goal of science. 176

12.2 Science and religion. 179

12.3 God's DNA. 183

12.4 Beyond DNA. 189

12.5 The Universe and God. 190

12.6 Closing Thoughts. 194

Simple Derivation of Special Relativity Using the Pythagorean Theorem 197

A.1 Simple calculation of space contraction and time dilatation. 198

Calculations for the Addition of Relative Velocities Considering the Effects of Special Relativity 203

B.1 Adding relativistic velocities. 204

B.2 Adding everyday velocities. 209

B.3 The Lorentz transformations. 210

B.4 Deriving the Relativistic Velocity Addition Formula. 211

B.5 Considerations for General Relativity. 215

How to Remember Relativity Principles....................217

C.1 Reviewing the pieces together. 218

C.2 Exercises that can turn everyday experience into relativity classrooms. 221

References and Sources for Further Reading.................... 229

Index 231

List of Photos & Illustrations

What Is Light?

Figure 2-1. Light is made of particles but moves in a wave-like fashion. 16

Figure 2-2. Large distances are measured in terms of how long it takes light to traverse them. 18

Figure 2-3. The speed of light is always the same. 19

Figure 2-4. Light is both a wave and a particle. 20

Figure 2-5. The color of light is determined by its wave characteristics. 21

Figure 2-6. Light is the visible portion of the larger electromagnetic spectrum. 22

There Is No Up or Down in Space or Anywhere Else

Figure 3-1. Before Einstein, many scientists believed there was an invisible structure that defined space called the aether. 25

Figure 3-2. Einstein developed his theories at a time before widespread use of electricity. 27

Figure 3-3. We judge motion relative to our surroundings. 28

Figure 3-4. Motion of different platforms can be combined (added). 29

Figure 3-5. Speed depends on the object chosen as reference. 29

Figure 3-6. Our movements against some references are huge. 30

Figure 3-7. Many frames-of-reference are available for measuring motion. 31

Figure 3-8. We can measure the speed of motion relative to the train car. 32

Figure 3-9. We can calculate the speed of motion relative to the Earth. 33

The Special Theory of Relativity

Figure 4-1. The speed of light is measured within a train car. 37

Figure 4-2. Light appears to travel a shorter distance on the train compared to as seen from the ground. 38

Figure 4-3. Since the speed of light is always the same, then time must go slower on moving platforms. 39

Figure 4-4. The speed of light is measured on the ground. 42

Figure 4-5. The clock that appears to run slow depends on the frame-of-reference from which measurements are taken. 43

Clocks

Figure 5-1. Clocks have three main components: the display, power source and oscillator. 47

Figure 5-2. How a light-clock works. 48

Figure 5-3. A light-clock can illustrate time slowing on a moving object. 50

Time and Space

Figure 6-1. How high-speed travel affects the passage of time for the "moving" rocket ship. 56

Figure 6-2. How high-speed travel affects the passage of time for the "moving" Earth / star. 61

Shortcomings of Special Relativity and the Equivalence Principle

Figure 7-1. Is the bucket spinning, or is the Universe spinning around the bucket? 65

Figure 7-2. Is the rocket ship accelerating through the Universe, or is the Universe accelerating past the rocket ship? 66

Figure 7-3. Spinning objects in weightless space experience centrifugal forces. 67

Figure 7-4. Interaction of centrifugal force and gravity. 68

Figure 7-5. Gravity and acceleration are equivalent. 70

Figure 7-6. A more massive object is harder to accelerate. 71

Figure 7-7. All objects fall at the same rate. 72

Figure 7-8. Einstein combines gravity with acceleration. 73

General Relativity: Reality in Four Dimensions

Figure 8-1. Adding dimensions affects geometry in unexpected ways. 82

Figure 8-2. Seeing space and time as an integrated entity. 83

Figure 8-3. How gravity distorts space-time made obvious in a three-dimensional analogy. 86

Figure 8-4. How the bending of space-time explains the otherwise mysterious "pull of gravity." 88

Figure 8-5. How the bending of space-time even effects the movement of light. 88

Figure 8-6. Free-fall means following the natural contours of space-time. 91

Figure 8-7. How gravity and acceleration feel. 92

Figure 8-8. Accelerations created by accelerating objects and those due to gravitational fields feel different. 93

Figure 8-9. The shapes of gravitational fields can vary. 96

Figure 8-10. How gravity and acceleration affect the movement of light. 99

Figure 8-11. Why our normal Euclidian geometry doesn't work for relativity. 100

Figure 8-12. Like waves on a pond, the frequency of light changes with relative motion. 103

Figure 8-13. The affect of relative motion, acceleration and gravity on the color of light. 105

Figure 8-14. Large gravity-producing bodies distort space-time in a way that causes space to be compressed and time slowed. 108

What Relativity Says About the Universe: Part 1 — Time, Space and Aging

Figure 9-1. Einstein's twin paradox. 111

Figure 9-2. The twin paradox — Earth-bound perspective. 113

Figure 9-3. The twin paradox — spaceship perspective. 113

Figure 9-4. Adding velocities makes common sense. 115

Figure 9-5. Adding velocities considering the effects of relativity (nested example). 116

Figure 9-6. Adding velocities considering the effects of relativity (approaching example). 117

Figure 9-7. Adding velocities of photons considering the effects of relativity. 118

Figure 9-8. Time stops and space compresses to zero at the speed of light. 119

Figure 9-9. Why simultaneous events aren't simultaneous on moving platforms. 122

Figure 9-10. Relative motion can change the order that events happen. 123

Figure 9-11. Relative motion cannot change the order of events when they are causally related. 124

Figure 9-12. Passing trains define events for analyzing the effects of Special Relativity. 126

Figure 9-13. Relative motion changes the order of events and makes slowed time and compressed space obvious (Train 1, Part A). 127

Figure 9-14. Relative motion changes the order of events and makes slowed time and compressed space obvious (Train 1, Part B). 127

Figure 9-15. Relative motion changes the order of events and makes slowed time and compressed space obvious (Train 2, Part A). 128

Figure 9-16. Relative motion changes the order of events and makes slowed time and compressed space obvious (Train 2, Part B). 128

What Relativity Says About the Universe: Part 2 — How the Universe is Made

Figure 10-1. Locations are defined by a spatial coordinate and a time coordinate. 132

Figure 10-2. Space-time diagram with example of possible travel. 134

Figure 10-3. Space-time diagram for the twin paradox. 135

Figure 10-4. Space-time diagrams establish limits on information communication. 136

Figure 10-5. Space-time diagram for communication with Mars. 137

Figure 10-6. Distance to the Andromeda Galaxy. 138

Figure 10-7. Photons experience complete time slowing and space compression. 140

Figure 10-8. The Universe is expanding. 142

Figure 10-9. The Universe is expanding in four dimensions. 143

Figure 10-10. The double slit experiment. 146

Figure 10-11. Time slowing and space compression result from changes in viewing perspective. 149

Figure 10-12. How magnetic fields are created. 152

Figure 10-13. How gravity waves work. 153

Where's the Proof?

Figure 11-1. At first, most scientists didn't accept Einstein's theories. 163

Figure 11-2. Eventually, Einstein was congratulated for his work. 164

Figure 11-3. The orbit of Mercury exhibits a small shift due to the bending of space-time near the Sun. 166

Figure 11-4. Deformed space near large massive bodies deflects the path of light. 167

Figure 11-5. Deformed space near large, massive bodies can create multiple images of a single object. 167

Figure 11-6. The frequency of light coming from the Sun has been shown to decrease as it reaches Earth. 168

Figure 11-7. Time slowing on moving clocks has been demonstrated. 169

Figure 11-8. The decay rate of high-energy radioactive particles moving at near the speed of light is slower due to Special Relativity. 170

Figure 11-9. Space-time compression closer to Earth's surface was demonstrated. 171

Figure 11-10. LIGO Gravity Wave Detector. 172

Figure 11-11. Modern GPS satellites apply corrections to their clocks to account for the less compressed space-time at a higher altitude. 174

Science and Religion

Figure 12-1. The early stages of Universe evolution. 185

Figure 12-2. The evolution of life. 188

Figure 12-3. The evolution of information beyond biology. 190

Simple Derivation of Special Relativity Using the Pythagorean Theorem

Figure A-1. The Pythagorean theorem. 198

Figure A-2. The Pythagorean theorem is used to calculate the amount of time slowing and space contraction (i.e., the relativistic correction factor). 199

Calculations for the Addition of Relative Velocities Considering the Effects of Special Relativity

Figure B-1. Adding velocities considering the effects of relativity (nested example). 205

Figure B-2. Adding velocities considering the effects of relativity (approaching example). 207

Figure B-3. Photons approaching other photons. 208

Figure B-4. Adding velocities for normal speeds considering the effects of relativity (nested example). 209

How to Remember Relativity Principles

Figure C-1. Integrated view of Special Relativity. 219

Figure C-2. Integrated view of Special and General Relativity. 220

Figure C-3. Imagining the effects of Special Relativity (high speed travel). 222

Figure C-4. Imagining the effects of Special Relativity (moderate speed travel). 223

Figure C-5. Imagining the effects of Special Relativity (without relative motion). 224

Figure C-6. Exaggerated effect of General Relativity when in a tall building. 225

Figure C-7. Exaggerated effect of General Relativity when in an accelerating elevator. 226

Chapter 1
Introduction

This chapter addresses the following topics and questions:

1.1 Overview. *How is relativity addressed in this book?*

1.2 Target audience. *Who should read this book?*

1.3 Pondering framework. *How is this book designed to present relativity in a way that makes it understandable?*

1.4 What you will learn. *What does this book cover?*

When first introduced by physicist Albert Einstein in 1905 and later expanded in 1915, the theories of relativity were controversial and not universally accepted. Since then, many scientific studies have confirmed the theories' predictions. Today, relativity is central to our understanding of the Universe. This book introduces the concepts of relativity for the non-scientist using coordinated illustrations that help visualize how relativity works. The illustrations are sequenced throughout the book in a way that helps build and reinforce understanding gradually.

1.1 Overview. This book describes relativity in the simplest of terms with illustrations to help visualize and understand its principles.

The concepts of relativity are not as difficult to understand and appreciate as traditional thinking would have us believe. In fact, the principles of relativity can be understood at a conceptual level without any math at all. And high school algebra can suffice for exploring the basic workings of Special Relativity, the first and easiest of Einstein's theories of relativity. The biggest difficulty in understanding relativity is that its effects only become obvious under conditions not normally experienced in everyday life. This makes it hard to appreciate how relativity works. It is therefore beneficial to study relativity using imagined situations that expand everyday experiences to incorporate conditions under which the effects of relativity are evident and can be envisioned.

This book aims to do that, building an understanding of relativity using illustrations and examples that allow the reader to envision the conditions under which relativity can be experienced. It presents the basic concepts of relativity beginning with the observational foundations that drove Einstein to propose it. It makes the case for Einstein's initial theory of relativity, called the Special Theory of Relativity, by showing how it resolves inconsistencies between how we traditionally measure motion and the measurements made of the speed of light. The book then builds on this foundation to define and explain the more complete and comprehensive theory of relativity, called the General Theory of Relativity. General Relativity expands Einstein's initial theory to also explain how acceleration and gravity affect the shape of space and the passage of time. After describing and depicting the Special and

General Theories of Relativity, implications for how the Universe works are explored. Mental exercises and illustrations for remembering and internalizing the principles of relativity are presented in Appendix C.

As noted, this book uses diagrams to graphically show how relativity works. The diagrams work together to gradually build an appreciation of relativity principles. They help develop an internal visual picture of relativity and thus make relativity easier to understand. However, while the diagrams help with visualizing and understanding relativity, they are only a visual approximation to how relativity actually works. The diagrams depict relativity principles on the flat, two-dimensional pages of this book, while relativity really works in a four-dimensional space-time. Four-dimensional space-time is what Einstein says results from combining three-dimensional space (that has three directions you can go: forward-back, right-left, and up-down) with time. This is an important part of relativity and is discussed in detail in this book. The simple two-dimensional diagrams help with visualizing and understanding how relativity works, while the actual four-dimensional geometries are too complex to accurately represent in pictures.

1.2 Target audience. This book is written for anyone who would like to understand the principles of relativity. A basic understanding of high school math is helpful but not necessary.

This book is written for non-scientists. The concepts are presented in their simplest form with pictures and illustrations to help achieve understanding. No math is needed, although a few simple formulas are presented for those interested, just to give deeper insights into the workings of relativity. For example, simple algebraic formulas are provided to show why it is not possible to travel faster than the speed of light. It is not necessary to understand the formulas in order to appreciate the basic concepts of relativity. The reader is able to skip them entirely, if desired.

There is no reason that anyone who has an interest should avoid learning about relativity. Relativity is something that anyone with a high school education can understand, at least at a conceptual level, and enjoy. Illustrations and simplified descriptions are provided to help grasp otherwise complex principles that define relativity. The goal is to develop an understanding of the concepts of relativity, building from basic human experience and common sense. For those interested in a more in-depth discussion of relativity, references for further reading are provided. Also Appendices A and B provide some very basic (high school level) mathematical descriptions and Appendix C includes exercises to help remember the concepts.

1.3 Pondering framework. The concepts and principles of relativity are presented using illustrative figures, consistent perspectives and simple graphics. This approach facilitates visualizing and learning, and supports ongoing pondering as a way to build a deeper and more enduring understanding of relativity.

This book is a "pondering book." It is meant to develop understanding through a "pondering" process in which the concepts of relativity are internalized by ongoing and repeated visualization. This is important because the concepts of relativity require adopting a new way of seeing the world that (initially, at least) seems inconsistent with everyday experience. This seeming inconsistency is because we experience the world from an Earth-bound human perspective. The human experience does not include near light speed motion, under which the workings of relativity would be self-evident. In order to provide a perspective where the effects of relativity can be shown, this book extrapolates from common everyday situations to imagine conditions under which the principles of relativity can be visualized. Numerous illustrations with explanations are provided to help in this process.

Finally, this book presents the concepts and foundations of relativity from a learning perspective. Material is organized and presented in a logically structured way that builds an understanding of foundational concepts before developing more difficult ones. The illustrations reinforce the material and help readers visualize and internalize the concepts, building on our human-scale experience and backgrounds. Simple coordinated figures are provided throughout the book as an aid for building understanding in a logical sequence. Each chapter begins with a list and summary of major points covered therein.

1.4 What you will learn. The contents of this book are organized for gradual learning and building an understanding of relativity.

Table 1-1 provides an overview of the contents of this book. It summarizes the main points covered in each chapter. It also offers a good review of the material and a reference for remembering the larger context of material presented. Readers can refer back to this table as needed to see the larger context and review prior material, and to support the pondering process.

Table 1-1. Overview of the Contents of this Book

1. Introduction	- **Overview.** This book describes Special and General Relativity in simple terms with many illustrations to help visualize and understand the concepts. - **Target audience.** This book is written for anyone who would like to understand the principles of relativity. A basic understanding of high school math is helpful but not required. - **Pondering framework.** The concepts and principles of relativity are presented using graphics and visual components, building from consistent perspectives. This approach facilitates learning and supports ongoing pondering as a way to develop a deeper and more enduring understanding. - **What you will learn.** The main points of each chapter of this book are summarized. This summary introduces the organization and contents of this book. It also provides a contextual foundation for structuring your understanding of relativity as a tool for subsequent review.
2. What is Light?	- **The nature of light.** Knowledge of how light behaves is central for understanding relativity. This chapter describes important characteristics of light. It describes what light is made of and presents key aspects of light's behavior that form the foundation for understanding relativity. - **The size of the Universe.** This section describes how we use light to measure the size of the Universe. - **Unexpected observations.** Some unexpected discoveries about light that led Einstein to develop his theories of relativity are discussed. - **The color of light.** This section describes how the color of light is determined. We can tell a lot about the effects of relativity from its effects on the color of light. - **The electromagnetic radiation spectrum.** This section explains how light is really just a portion of the larger spectrum of electromagnetic radiation, including radiation associated, for example, with radar, radio, and x-rays.

Table 1-1. Overview of the Contents of this Book - *Continued*

Chapter	Contents
3. There is No Up or Down in Space, or Anywhere Else	- **The aether.** Many scientists before and at the time of Einstein believed that there was an invisible structure that defined space and against which all motion could be judged and measured. This structure was called the "aether." The acceptance of relativity led scientists to abandon the idea that there is a space-defining aether. - **The nature of motion.** Motion of any object can only be defined and measured relative to the position and motion of other objects. This is called Galilean Relativity because it was the Italian Renaissance scientist Galileo Galilei who first proposed it. Galilean Relativity forms a central principle underlying Einstein's Special Theory of Relativity. - **Incompatibility between relativity and the aether.** Einstein's theories of relativity were incompatible with the existence of an aether. This disagreement put Einstein at odds with many of his colleagues. The eventual acceptance of Einstein's theories ultimately led to the abandonment of belief in a space-defining aether. - **The velocity of motion is relative too.** The understanding that an object's motion must be measured relative to other objects led Einstein to realize that velocity (the time-distance relationship between objects) depends on what object is chosen as a reference for measuring the motion. This formed the foundation for the Special Theory of Relativity.
4. The Special Theory of Relativity	- **How Special Relativity works.** This section describes how the behavior of light and Galilean Relativity combine to form the basis of Einstein's Special Theory of Relativity. Illustrations show how the Special Theory of Relativity provides a new way of understanding time and space. It graphically shows why under Special Relativity, time slows and space compresses when objects move "uniformly" relative to one another, and why "moving clocks run slow." - **Why Special Relativity is called "special."** This section answers the question, what is so "special" about Special Relativity? - **Observer perspective is the key.** The perspective of the observers in considering properties of moving objects is central in defining the effects of Special Relativity. This section explains why.
5. Clocks	- **Clocks.** Clocks are central to understanding relativity. This chapter asks the questions: what is a clock and how are clocks related to time and relativity? - **Light-clocks.** Hypothetical clocks called light-clocks use the movement of light to measure time and provide significant insight into how relativity works. Light-clocks are introduced in this chapter and used throughout the rest of the book to make relativity understandable. - **Clocks and time.** Questions of whether hypothetical light-clocks correlate with the "real" clocks available to scientists and people are addressed in this section.

Table 1-1. Overview of the Contents of this Book - *Continued*

Chapter	Contents
6. Time and Space	- **Time slowing.** The relationship between relative motion - the movement of objects relative to each other - and the slowing of time as defined by Special Relativity is described more completely in this section. - **Space compression.** When time is slowed due to the effects of Special Relativity there is a commensurate compression of space that is experienced. This relationship is presented, explained and illustrated. - **Bi-directional application.** There is no preferred frame-of-reference when considering the effects of Special Relativity. For example, when two objects are moving relative to each other, either object can be considered to be the fixed reference for measuring Special Relativity's effects.
7. Shortcomings of Special Relativity and the Equivalence Principle	- **Shortcomings of Special Relativity.** Accelerations, such as the centrifugal force in spinning objects, limit the applicability of Special Relativity and establish a beginning point for understanding General Relativity. This section describes why. - **The nature of acceleration and gravity.** This section describes how acceleration and gravity work together and why they are really the same force. - **The Equivalence Principle.** This section describes how "inertial mass" (that makes objects keep moving) and "gravitational mass" (that gives large objects like the Earth gravity) are equivalent and therefore why acceleration and gravity have similar effects on time and space.
8. General Relativity: Reality in Four Dimensions	- **Doubling down boldly.** The Special Theory of Relativity defined new ways of understanding space and time that went against and beyond previous conventions of understanding. But compared to General Relativity these advances were minor. Einstein was truly "doubling down boldly" when he proposed General Relativity. - **General Relativity requires a new kind of geometry.** The everyday mathematics and geometry we use to describe three-dimensional space are not adequate for describing General Relativity. This section explains how Einstein had to find a new mathematical geometry that could work for his four-dimensional Theory of General Relativity. - **Space and time do not exist separately.** One of the most fundamental outcomes of Einstein's theories of relativity is that space and time are not separate things. - **Space-time becomes distorted under conditions of gravitation.** Gravity affects how things move through space-time by affecting the shape of space-time. - **Acceleration and gravity are equivalent but different.** Gravity and acceleration are equivalent - they feel the same, are related to mass in a similar way, and can be combined like when banking a bicycle during a turn. However, they also have some important differences - being in an accelerating rocket ship is not the same as standing on Earth's surface. - **Motion and space-time distortions cause changes to light's appearance.** Relative motion changes the frequency (color) of light. Acceleration causes the color of light to keep changing as the speed of motion keeps changing. Gravity has a similar effect. Illustrations are included to help explain. - **Extreme gravitational conditions create extreme time slowing and space compression.** Under extreme gravitational conditions, time can slow to the point of stopping and space can become so compressed that light cannot escape its hold. This happens around "black holes" described in this section.

Table 1-1. Overview of the Contents of this Book *- Continued*

Chapter	Contents
9. What Relativity Says About the Universe: Part 1 - Time, Space and Aging	- **Time and aging.** Does relativity affect how fast we grow old? Einstein's twin paradox used a space-traveling example to answer this question. The twin paradox is presented in this section. - **The universal speed limit.** Why are relative speeds faster than the speed of light, called speed "c," not possible? This section shows why. - **Simultaneous events and causality.** Given the effect of relative motion on time, is there really such a thing as simultaneous events? Can relativity upset the order of events such that results can occur before the events that caused them? If not, why not? This section answers these questions with real world examples. - **The relationship between event ordering, the slowing of time and the compression of space.** This section graphically shows how relative motion that affects the ordering of events does this by compressing space and slowing time. It's obvious when you see it.
10. What Relativity Says About the Universe: Part 2 - How the Universe Is Made	- **Moving through space-time.** This section explores how the universal speed limit (speed "c") and the one-way nature of time (always going from past to future) combine to limit our movements within space-time. - **Relativity limits information access.** Just as important as the limits that relativity puts on our movements through space-time are limits relativity places on what we can know. This section shows why. - **Time it takes for light to travel across space.** Is it really true that it takes light many years to reach us from distant stars and galaxies? Well no, but why not? This section explains. - **Exceeding the universal speed limit.** Really, isn't there anything that can go faster than speed "c"? Well OK, yes there is, but what is it? This section describes the few situations in which faster than speed "c" movement is at least implied and may be possible. - **The appearance of compressed space or actual compressed space?** Do things actually become shorter with the compression of space or is it just the appearance of things (and space) shortening? The answer is both yes and no. This is described in this section. - **How objects bend space-time.** How is it that massive objects are able to bend space-time? How do things far away from the Earth know that the Earth is "pulling" on them? How does gravity extend across space-time? This section answers these questions. - **The structure of space-time.** This section explores the question: what is space-time made of? Is space-time really empty? - **$E=mc^2$.** Energy (E) and mass (m) are interchangeable as defined in Einstein's famous equation, $E=mc^2$. This famous interrelationship between mass and energy defines important characteristics of how the Universe is put together and works. So what does $E=mc^2$ have to do with relativity? This section explores the ramifications of Einstein's famous equation. - **Limits of relativity's scope and competence.** Are there limits to what relativity can tell us about the Universe? Yes, this section describes them.

Table 1-1. Overview of the Contents of this Book - *Continued*

Chapter	Contents
11. Where's the Proof?	- **The questioning scientific community.** This section describes how the scientific community responded to Einstein's theories: at first skeptical and eventually congratulatory. - **Verification of Einstein's predictions.** Einstein made three major predictions that he said would validate his theories. This section describes them. - **Modern experiments have measured the effects of relativity.** This section discusses the results of recent scientific studies conducted to test Einstein's theories. Measurements verifying relativity's predictions about the effects of relative motion and gravity on time and space are described. The very recent discovery and measurement of gravity waves moving through space are also presented. - **Putting relativity into practice.** Beyond theory and experiment, the proof is in the pudding. That is, does relativity work in the real world? This section describes how our knowledge of relativity makes modern technology and conveniences possible.
12. Science and Religion	- **The goal of science.** The work that scientists do along with the deeper meanings and perspectives that accrue from an understanding of science and relativity are discussed. - **Science and Religion.** How can our growing understanding of science inform our religious beliefs? How can religion inform and motivate scientific progress? What can dogma-based religions do to keep up with science? This section discusses these important questions and suggests a way forward. - **God's DNA.** Science teaches us that the Universe began in a giant "Big Bang" 13.8 billion years ago and has been expanding and evolving ever since. This section discusses that evolution, our growing understanding of how the Universe works, and asks the God question. - **Beyond DNA.** With the creation of humans, a new kind of evolution that occurs outside of biological structures has begun. - **The Universe and God.** Scientific discoveries about the Universe and its evolution invariably raise questions about Creation and God. This section discusses these questions in relation to science. - **Closing thoughts.** The main goals of this book are noted.

For those interested, three appendices are provided to give a deeper understanding of relativity. The first two appendices offer a glimpse at the mathematical underpinnings of Special Relativity. These are presented using high school level math for ease of understanding. The third appendix includes some mental exercises that can help in remembering the concepts of relativity. These appendices are summarized in Table 1-2.

Table 1-2. Overview of the Contents of the Appendices

	Contents
Appendix A. Pythagorean Illustration	- Special Relativity can be visualized at a basic level using the Pythagorean theorem. Appendix A provides this insight. It shows how a simple application of the Pythagorean theorem (from high school geometry) can be used to derive the formulas for time slowing and space compression as defined by the Special Theory of Relativity.
Appendix B. Calculations for the Addition of Relative Velocities	- Mathematical calculations for some of the examples from the book are provided for those interested. These examples use simple algebra to show why it is not possible to exceed the speed of light according to Special Relativity. - The Lorentz transformations used by Einstein to develop Special Relativity are described and related to the formulas for adding velocities across multiple frames-of-reference.
Appendix C. How to Remember	- The main lessons of this book are summarized to help in reviewing Special and General Relativity and for remembering the scientific contexts from which they were derived. - Exercises to help see how relativity works are presented as an aid for remembering the relativity concepts.

As noted, this book is organized so that earlier chapters provide a foundation for later material. Each chapter contains information that is expanded and built upon in later chapters. Graphical illustrations are included throughout. They are simplifications of the concepts intended to help in visualizing how relativity works. The graphic depictions and illustrations are repeated and expanded throughout the book as a way to help reinforce understanding. These illustrations are simple two-dimensional diagrams that help visualize how relativity works without trying to capture the deeper understanding that Einstein described with advanced mathematics and differential geometry.

Finally, it takes time to become comfortable with the concepts of relativity. You may follow all the arguments and logic as you read this book but still not feel entirely comfortable in your understanding of relativity. This is normal. Even professors that teach relativity admit that it differs from their common sense and normal experience, and is therefore difficult to fully appreciate at a human level. Nevertheless, with repeated pondering, the concepts of relativity will become more readily appreciated. You will gain a level of comfort with relativity as you read this book since the concepts are reinforced and reviewed throughout. Subsequent pondering will further build a more solid understanding.

Chapter 2
What Is Light?

This chapter addresses the following topics and questions:

2.1 The nature of light. *What is light made of and how does it move through space? What is it about light that is still not well understood?*

2.2 The size of the Universe. *How are the nature and speed of light used to measure the size of the Universe?*

2.3 Unexpected observations. *What unexpected discoveries about light led Einstein to develop his theories of relativity?*

2.4 The color of light. *How is the color of light determined?*

2.5 The electromagnetic radiation spectrum. *How is light related to other forms of electromagnetic radiation, like radio waves?*

The nature of light and how we experience and measure it are fundamental to understanding the theory of relativity. This chapter describes light and builds this foundation, which also provides important insights into the size of the Universe. It's also important to appreciate that light is just a part of the larger electromagnetic spectrum, which includes, for example, radio waves and x-rays. While it is the nature and behavior of light that is referred to throughout this book, the points made about light and its relation to relativity are also true for all forms of electromagnetic radiation.

2.1 The nature of light. Knowledge of how light behaves is key to understanding relativity. This section describes important characteristics of light. It describes what light is made of and how it moves through space. It describes how light is both a wave and a particle.

Light (and all forms of electromagnetic radiation) is a kind of energy that is composed of many trillions of miniscule particles called photons. Photons are extremely small, so small that they have no mass at all! For this reason, they are really more like tiny bundles of energy than actual particles. However, since photons have a specific location when measured and exhibit particle-like characteristics, such as spin, they can be thought of as particles, just extremely tiny.

Photons move along straight paths[1] but in a wave-like manner. This dual particle and wave behavior of light is not well understood. The wave-like nature of light is comparable to waves on a pond and sound waves. However, unlike waves of sound and water, light waves do not need a medium such as air or water to move through. Light is able to move through empty space.[2] Light propagates as a wave but at its most elemental level is made of particles (photons) with a measurable position and momentum, but no mass.

The discovery that light is composed of energy particles called photons but also behaves in a wave-like manner was unexpected. This schizophrenic behavior of light is called "wave-particle-duality." When looking at the

1 Light actually travels along many paths, not all straight, but only light traveling in straight-line paths has a high likelihood of reaching a destination. See *QED: The Strange Theory of Light and Matter* by Richard Feynman for a description, including a discussion of ways to trick light into traveling along paths that are not straight!

2 There is some question about whether space is really "empty." It is not made of normal matter like air or water and is therefore frequently thought of as being empty. This is discussed more fully later in this book.

particle aspect of light, scientists find that the photons can be detected as energy particles with a specific location (for example, when they strike an object such as photographic film). However, it is the light's wave properties that define important characteristics of the light, such as its color. And like waves on a lake, light waves can interact with each other forming interference patterns that combine to distort the appearance of the light. But the key to remember here is that light is both a wave and a particle.

This "wave-particle duality" still stirs uncertainty about the true nature of matter, energy, and the Universe today. In this book, light will be depicted showing both its wave and particle aspects (like Figure 2-1) to remind us that light is both a wave and a particle.

Because photons have no "rest mass"[3] they travel faster than anything else in the Universe. They move at the speed of light. Physicists determined that the movement of light between locations is not instantaneous. They estimated its speed as early as 1676 and first accurately measured its speed in the 1920s.[4] Today, we have very accurate measures of light speed. Light travels at 186,282 miles (299,792 kilometers) every second through a vacuum[5] (and slightly slower through air, water, and other medium[6]). Scientists use the letter "c" to represent the speed of light. At this speed, light would go around the Earth almost 7½ times in one second! No wonder the room gets light so fast when we turn on the light switch.

3 Scientists describe photons as having no mass, but having momentum. They therefore describe photons as having no "rest mass." That is, they have no mass at rest. However, photons always move at the speed of light. This gives them momentum that is manifested as energy.

4 In 1676, Danish astronomer Ole Romer determined that the movement of light between locations is not instantaneous. He estimated the speed of light by measuring when Jupiter's moon Io appeared from behind Jupiter, taking the measurements when Earth's distance from Jupiter varied. Physicist Albert Michelson, together with Scientist Edward Morley, used an interferometer developed by Michelson to measure the speed of light. Working at what is now Case Western Reserve University they were the first to accurately measure the speed of light in the 1920s. This famous experiment is now referred to as the Michelson-Morley experiment.

5 Scottish mathematical physicist, James Clerk Maxwell, developed equations for electromagnetic radiation in 1865 that predicted the speed of light. This prediction was later confirmed by experimental measurement.

6 Actually light slows by different amounts for different colors. This is how a prism works and why rainbows form. Throughout this book the speed of light is simply referred to as speed "c" and ignores the fact that its speed slows slightly when passing through things like water, glass and air.

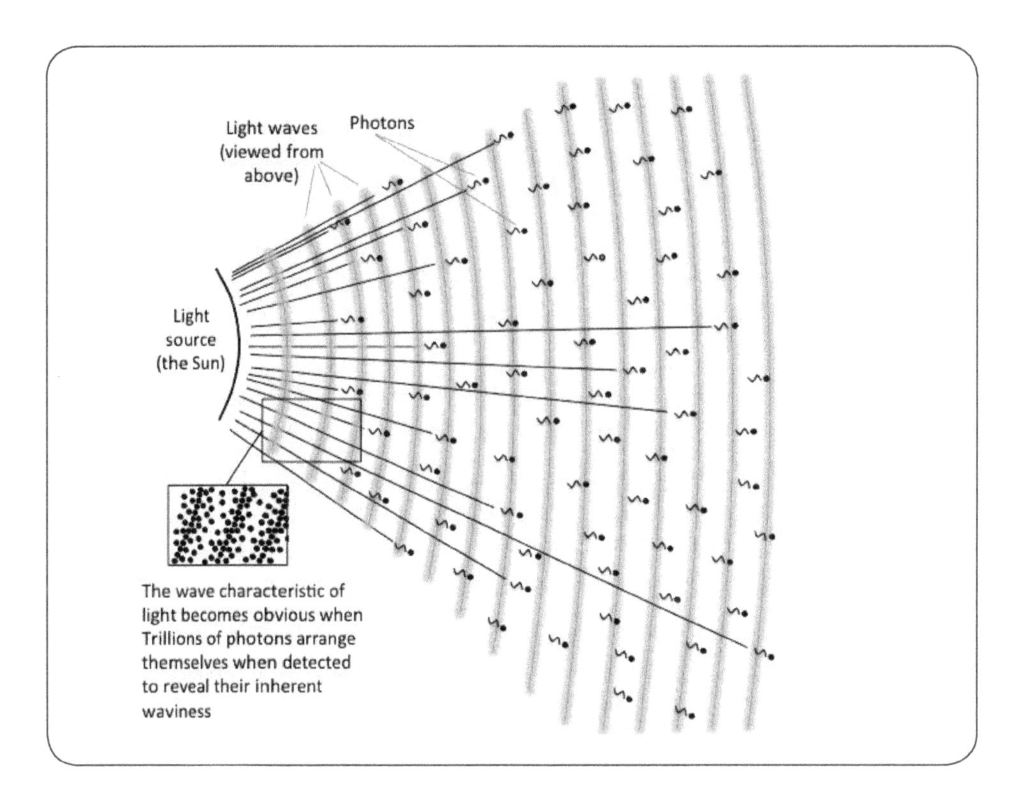

Figure 2-1. Light is made of particles but moves in a wave-like fashion. *Light is made of tiny particles (actually more like bundles of energy) called photons. Light also moves in a wave-like manner similar to how waves move on a pond, except that light can move through empty space. Photons move (mostly) along straight-line paths. Light's wave properties define the light's characteristics, such as its color.*

Even at this unimaginably high speed, it still takes light eight minutes to get to us from the Sun and 5½ hours to reach Pluto, which was until recently the furthest planet (now considered a dwarf planet because of its small size) from the Sun. It takes 11 hours for the Sun's light to reach Eris, now the furthest known (dwarf) planet. Eleven hours! And it takes over four years for light to reach us from the nearest star beyond our Sun, a multiple star system containing three stars called Proxima Centauri.

2.2 The size of the Universe. The Universe is so large that we measure distances across it in terms of how long it takes light to traverse it.

Our Sun is just one of billions of stars in our galaxy, the Milky Way. It is estimated that the Milky Way contains over 100 billion stars! The size of our galaxy and distances between galaxies is so large that they are measured by the amount of time it takes light to cross them. For example, it takes light 100,000 years to go across the Milky Way. And the Milky Way galaxy is just one of hundreds of billions of known galaxies. Light from the Andromeda Galaxy, the nearest galaxy beyond our own Milky Way, takes 2½ million (2,500,000) years to reach us[7]. There are at least 200 billion galaxies in the Universe, each containing over 100 billion stars on average. Because it takes light so long to get here from distant locations across the Universe, the further away we look the further back in time we are seeing.

The distance that light travels in one year is called a light-year. The light-year is the standard unit for measuring how far away things are within the Universe. Other more conventional measures of distance, such as miles or kilometers, are too limited for measuring the large distances across space. For example, if we were to use miles instead of light-years to measure the distance to even the nearest star, Proxima Centauri, it would be 4,178,304,000,000 miles. This is a very hard number to even say, let alone understand. It's easier to just say four light-years. And this also provides an intuitive appreciation of just how far the separations are. It takes light four years to reach us from the nearest star. The light we receive from Proxima Centauri is therefore four years old, which means that we are seeing Proxima Centauri as it was four years ago. This is described in more detail later in the book.

Another advantage of the light-year is that it captures both the time and space dimensions of the Universe. As discussed later, an important outgrowth of Einstein's theories is that time and space are really integrated aspects of a four-dimensional reality in which space and time do not exist separately but only in combination. They are integrated components of a four-dimensional space-time. The three dimensions of space (length, width and height) combine with time to form four-dimensional space-time. This is discussed in much more detail later in this book. For now, the point is that the light-year offers a better measure of distance through four-dimensional space-time than conventional measures of distance, such as miles. It is for these reasons that light-years (or light-hours or light-minutes for objects that are too close to be expressed in light-years) are used for describing distances

---

7 We will see later that it is not as simple as saying that light takes 2.5 million years to get to us from Andromeda because traveling at the speed of light affects time and space. It is more accurate to say that the light that reaches us from Andromeda left Andromeda 2.5 million years ago in our time reference.

across space-time. Figure 2-2 shows the distances to various locations in the Universe expressed in terms of the time it takes light to travel across it.

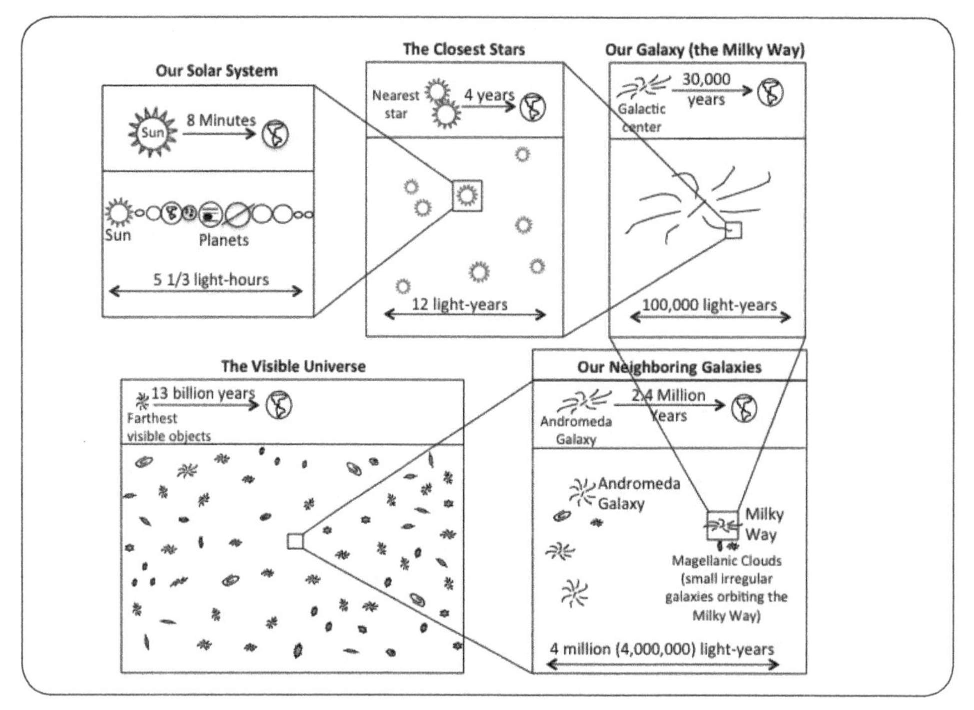

Figure 2-2. Large distances are measured in terms of how long it takes light to traverse them. *It takes light many, many years to reach us from distant galaxies and even takes eight minutes to get here from the Sun. The distances from Earth to different points in the Universe are shown here measured in terms of how long it takes light to travel between them and us.*

2.3 Unexpected observations. Some unexpected discoveries about light led Einstein to develop his theories of relativity.

What was surprising to scientists, and is critical to understanding relativity, is that when physicists measure the speed of light they find that it always travels at the same speed, referred to as speed "c" (186,282 miles per second). This is true regardless of whether you are moving toward the light source or away from it! It is difficult to understand why the Universe works this way; but we know that it does based on the measurements scientists have made. They discovered this by measuring the speed of the light coming from a nearby star, both at a point in Earth's orbit when Earth was moving toward the star and at a point when Earth was moving away from the star. Figure 2-3 graphically shows this experiment. This result has been replicated and verified in many other analogous situations, such as measuring the light from stars in double star systems that move toward us during one portion of their orbit and away from us during another part of their orbit. Light always travels at the same speed regardless of the relative movement of the light source.

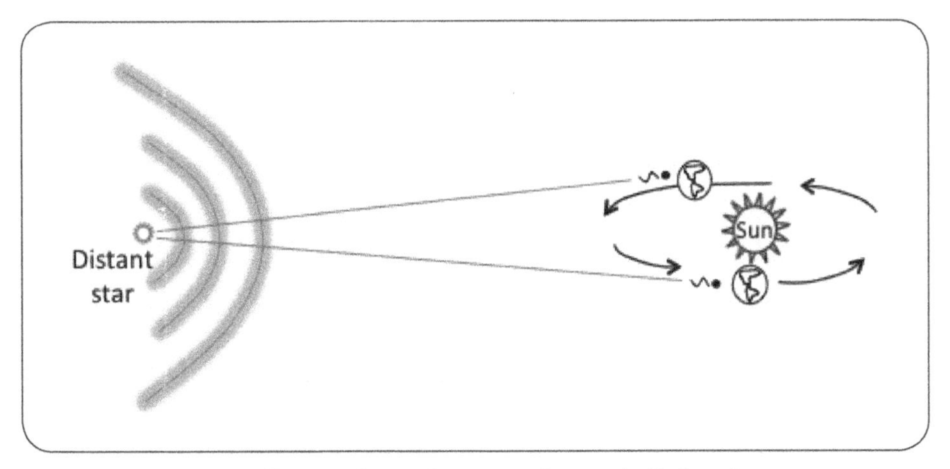

Figure 2-3. The speed of light is always the same. *The speed of light is the same no matter whether you are moving toward the light or away from it.*

This previously unexpected behavior of light puzzled scientists for many years. Why doesn't light move faster when we are moving toward the light source and why doesn't it move more slowly when we are moving away from it? While the underlying reasons for this are difficult to visualize, Einstein's theories of relativity provide a way of describing the Universe based on this behavior of light and in a way that is now supported by a large body of experimental evidence. In fact, relativity offers an explanation for this behavior, and also accurately predicted many other physical phenomena such as the bending of light near large massive stars. For these reasons Einstein's theories of relativity have become an accepted foundation for understanding how the Universe works. We'll see how relativity provides a good description of the Universe and the behavior of light (and of all forms of electromagnetic radiation) in subsequent chapters.

2.4 The color of light. An important characteristic of light is its color, which is determined by the spacing between the light waves.

As noted earlier, light (and all electromagnetic radiation) always travels at speed "c" (through a vacuum). However, light isn't always the same color. The color of light is determined by how closely the light waves are spaced. This is called the light's frequency. When the waves are spaced close together it has a higher frequency and the color of the light is toward the blue end of the color spectrum. When the waves are further apart, light has a lower frequency, and the color is more toward the red end of the spectrum. This is shown in Figures 2-4 and 2-5.

Like waves on the water, when you look at waves from the side they appear to oscillate up and down in a repetitive manner. Figure 2-5 shows this view for light waves and indicates how different frequencies are associated with different light colors. You can see this spectrum of colors when you look at a rainbow. A rainbow is created when water droplets in the air separate the Sun's light into its constituent colors. Red light has less closely spaced waves than blue light. Green and yellow light has wave spacing between that for red and blue. When all the colors of light are mixed together, the light appears white in color.

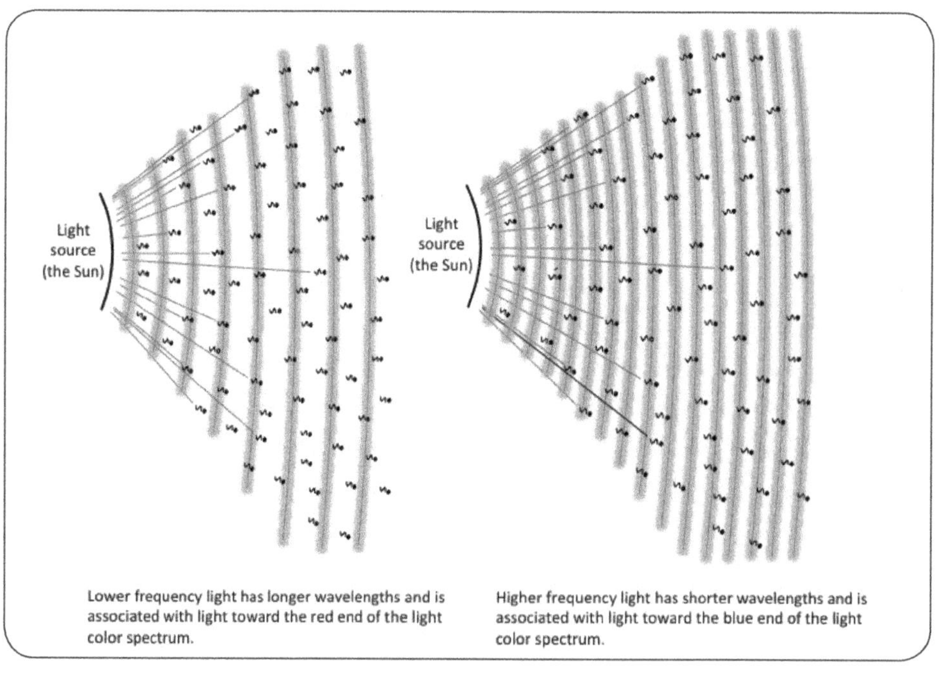

Figure 2-4. Light is both a wave and a particle. *It is composed of an energy particle and a wave. Light waves vary in frequency, or how closely the waves are spaced. The frequency of the light waves determines its color.*

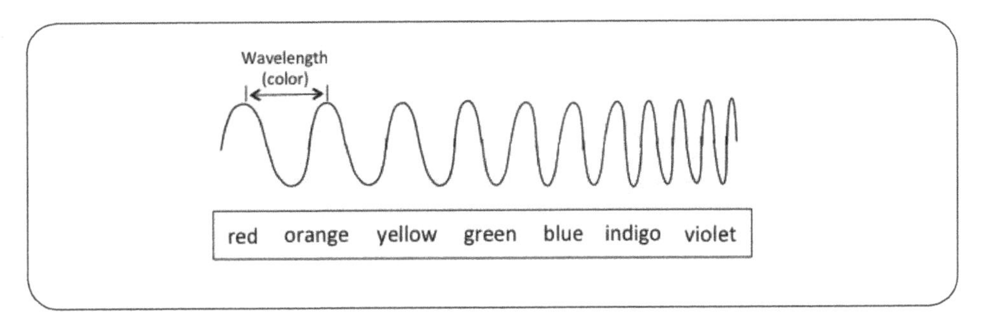

Figure 2-5. The color of light is determined by its wave characteristics. *Light waves drawn from the side appear to oscillate up and down like waves on a pond. The spacing or frequency of the waves determines the light's color.*

2.5 The electromagnetic radiation spectrum. Light is a form of electromagnetic radiation that covers a larger range of frequencies than just visible light. It is part of the larger electromagnetic radiation spectrum.

Light is just a portion of the larger electromagnetic radiation spectrum. The relationship of light to other parts of the electromagnetic radiation spectrum is shown in Figure 2-6. As shown in the figure, for example, radar and radio waves have lower frequencies while x-rays and gamma waves have higher frequencies. All electromagnetic radiation is characterized as both particles in the form of photons and as waves. It is the wave aspect that determines whether electromagnetic radiation is in the visible range (light) or outside of what we can see. We are only able to see electromagnetic radiation that is in the visible range, the part we call light. Scientists can see the other parts of the electromagnetic radiation spectrum using special instruments.

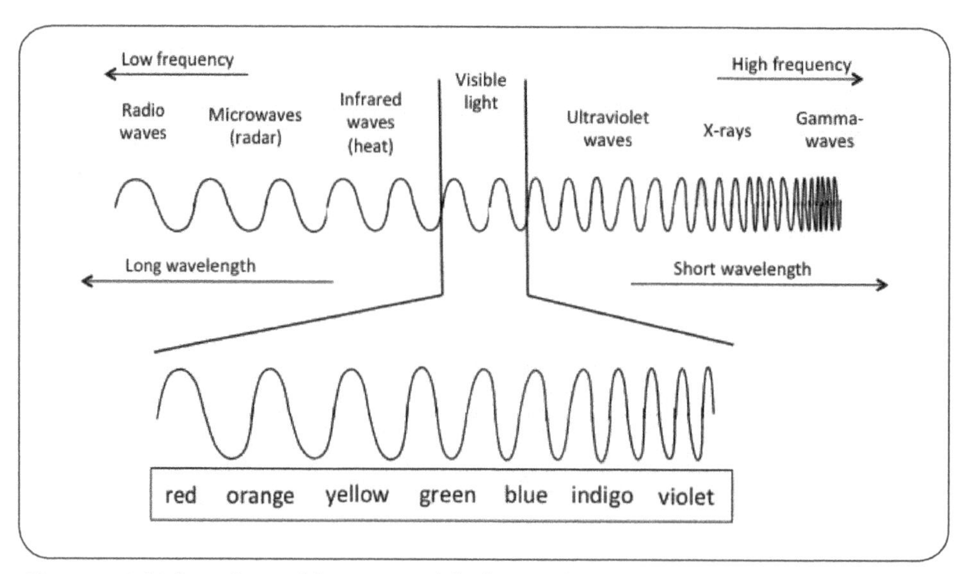

Figure 2-6. Light is the visible portion of the larger electromagnetic spectrum. *Electromagnetic waves that are in the visible range can be seen by humans and appear as different colors of light. Electromagnetic waves outside of the visible range cannot be seen by humans and go by other names, such as radio waves.*

Chapter 3
There Is No Up or Down in Space or Anywhere Else

This chapter addresses the following topics and questions:

3.1 The aether. *What did scientists think space was made of before Einstein? How did Einstein's theories of relativity give a different understanding that was more consistent with the behavior of light?*

3.2 The nature of motion. *What does the fact that the movement of an object can only be understood and measured relative to other objects say about the nature of motion?*

3.3 Incompatibility between relativity and the aether. *How were Einstein's theories of relativity incompatible with prior theories of the Universe that presupposed the existence of an aether against which all motion could be judged and measured?*

3.4 The velocity of motion is relative too. *How is an object's speed dependent on the frame-of-reference selected?*

Physicists tried to figure out why the measured speed of light was not affected by the motion of observers relative to the light source. This was a difficult task given the beliefs at the time about how the Universe was made.

3.1 The aether. Many scientists at the time of Einstein believed that there was an invisible structure that defined space and through which all objects moved. They believed that all motion could be judged and measured in comparison to this structure. This hypothetical structure was called the "aether" (pronounced ē-ther).

Belief among scientists that there had to be a structure defining space, a structure against which all motion could be measured and through which light was able to travel, was prominent at the time Einstein proposed his initial ideas about relativity. Much effort was expended trying to find this space-defining structure. This hypothesized structure was given a name. It was called the "aether." The aether was believed to exist everywhere and to provide an invisible yet real structure for all of space. The concept of an aether is simplistically illustrated in Figure 3-1 as a grid. However, repeated efforts to find the aether were unsuccessful. Scientists were unable to find any physical structure that made up space.

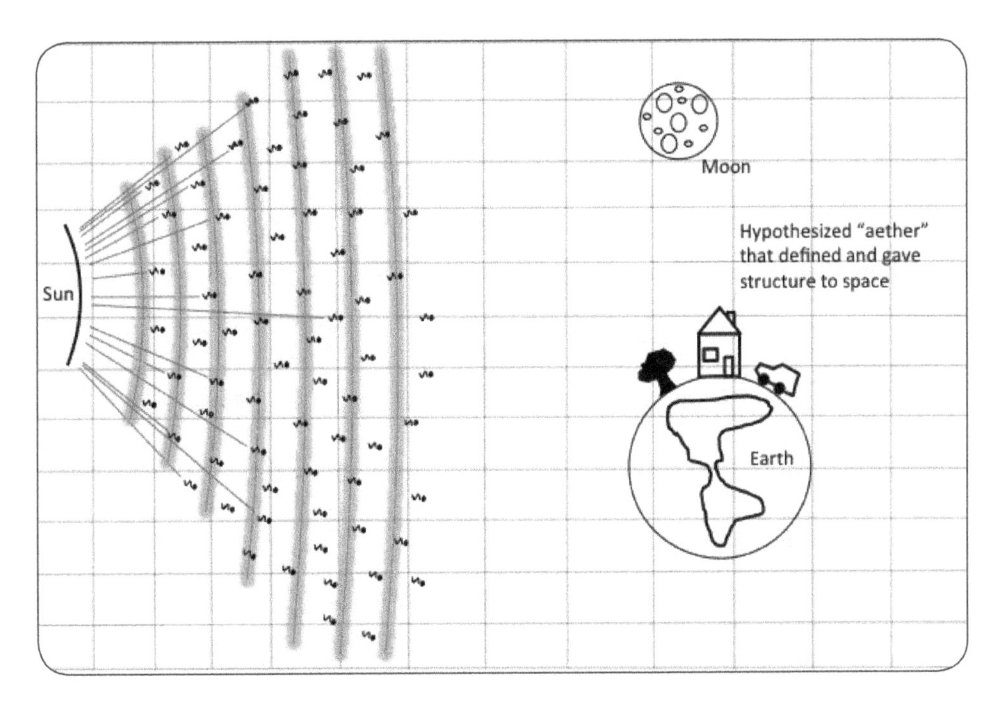

Figure 3-1. Before Einstein, many scientists believed there was an invisible structure that defined space called the aether. *It was believed that, once found, the aether would serve as a common reference for measuring all motion. It was also believed that the aether, represented as a grid in this illustration, was the medium through which light waves traveled. Today, we know that there is no aether defining space.*

The discovery that light always travels at the same speed regardless of any movement relative to the light source made the existence of an aether difficult to defend. This is because if space was defined by an aether that objects and light moved through, then light would move faster when we are moving toward the light source and slower when moving away. But this isn't what was found. Light always moved at the same speed, speed "c". Numerous hypotheses were developed to explain this discrepancy. For example, some suggested that the Earth distorted the aether or pulled a portion of it along as it moved, thus changing the apparent speed of light. But why would the speed of light always be exactly the same regardless of the direction the light was coming from? Many experiments were conducted to find evidence of an aether and to explain the unvarying speed of light. None were successful. Ultimately, it was the inability to find any evidence of an aether together with the acceptance of Einstein's theories for explaining light's unexpected behavior that ultimately put an end to the idea that there was an aether.

It was 1905 when Albert Einstein proposed the first of his two theories of relativity (a theory that he and some other physicists wanted to call the

Theory of Invariance because it explained the invariance of the speed of light). The term relativity was chosen because it reflects the underlying reality that all motion is relative. Today, Einstein's first theory is called the Special Theory of Relativity. The Special Theory of Relativity is described later in this book and still holds a central place in our understanding of how the Universe works. As we will see later, it was called "special" because it only applied to specific circumstances (i.e., special cases). Einstein's later theory of relativity was more general and covered a wider range of conditions. This broader theory is called the General Theory of Relativity.

It is interesting to note that when Einstein developed his Special Theory of Relativity in 1905, the world was still scientifically and technologically unsophisticated. For example, a significant portion of transportation was still accomplished using horse-drawn vehicles and electricity was just becoming available for personal and commercial use. There were no computers, except the human kind that found employment doing mathematical calculations by hand for a living.[1] Einstein's achievement in envisioning a new and novel way of describing the Universe was even more impressive given the times in which he lived, shown in Figure 3-2.

While Einstein's theories disproved the idea that there is an invisible structure defining space called the aether, this is not to say that there isn't a structure that defines the Universe. Whatever the structural nature of the Universe may be, we now know, based on Einstein's work, that it is four-dimensional, encompassing the three dimensions of space integrated with the fourth dimension of time. Einstein called it space-time, which will be discussed later in the book.

3.2 The nature of motion. Motion of any object can only be defined and measured relative to the position and motion of other objects. This is called Galilean Relativity because it was the Italian Renaissance scientist Galileo Galilei who first proposed it. Galilean Relativity forms a central principle underlying Einstein's Special Theory of Relativity.

1 In 1905, organizations employed people to perform tedious calculations in support of their operations such as adding numbers on a ledger. These workers were called "computers" because they performed computations for a living.

Figure 3-2. Einstein developed his theories at a time before widespread use of electricity. *Einstein's Special Theory of Relativity was proposed in 1905, a time when a significant portion of transportation was still horse-drawn and electricity was just beginning to be used as evident from this picture of Buffalo, NY taken in 1901.*

The basis for Einstein's new theories of relativity, in addition to the strange and unexplained behavior of light, was the understanding that an object's movement through space can only be defined and measured relative to other objects. The amount of any measured movement is defined by the extent of position change relative to a selected reference object. The amount and direction of an object's movement is completely dependent on what object is selected as reference. This is to say, there is no common or universal reference in space for measuring movement. Movement is always defined by position changes among selected objects. And this means that there is no aether; no universal structure defining space against which to measure motion.

We sometimes consider the Earth as the common reference for measuring motion because it is the center of our experience; but this is not adequate for movement far from Earth or even for all movement on or near Earth. Movement of any object, especially beyond the Earth, can only be measured in comparison to the relative positions of other selected objects. For example, the movement through space can only be measured in relation to objects in space that are selected, such as the Sun, Earth, or other planets. Even on Earth, the movement of a car can be measured in comparison to objects along the street (e.g., buildings) or in relation to other cars. The object selected for comparison is called the "frame-of-reference."

As noted earlier, this idea that all motion can only be defined in relation to other objects was proposed long before Einstein but was controversial because of the ongoing belief in an aether. The following discussion provides a graphical illustration of the relative nature of movement using common motions on Earth, and then relating them to the Earth's movement through space.

Consider a passenger walking down the aisle of an airplane forward toward the cockpit. He or she would be moving 1 or 2 miles per hour (mph) relative to the seats and walls of the plane, as shown in Figure 3-3. In this case, the plane would be the selected frame-of-reference.

Figure 3-3. We judge motion relative to our surroundings. *Walking down the aisle of an airplane, a passenger's speed is about 2 miles per hour compared to the inside of the plane.*

But if we also consider the movement of the plane over the Earth together with the passenger's forward movement on the plane, we would measure the speed as some 502 miles per hour relative to the Earth, assuming the plane is moving 500 miles per hour (i.e., the speed of the plane plus the speed walking down the aisle). This is shown in Figure 3-4. If the passenger were walking toward the rear of the plane, the combined speed would be 498 miles per hour (speed of the plane minus the walking speed).

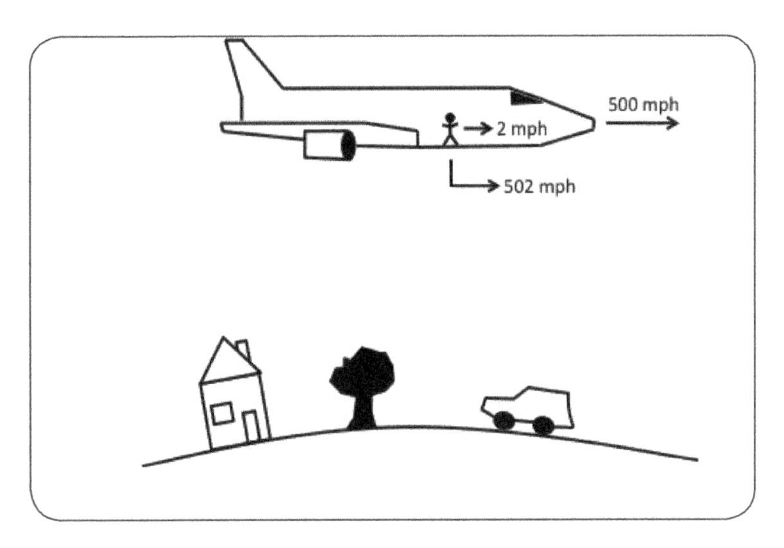

Figure 3-4. Motion of different platforms can be combined (added). *The passenger's speed on an airplane compared to the Earth is about 502 miles per hour, combining the speed of walking down the aisle toward the cockpit with the speed of the plane over the Earth.*

And if you consider the Earth's orbit around the Sun, with the Sun as the frame-of-reference, the passenger walking toward the front of the plane would be going 67,502 miles per hour (mph) relative to the Sun (i.e., the Earth's motion, about 67,000 mph, plus the movement of the plane, 500 mph, and passenger, 2 mph is 67,000 + 500 + 2 = 67,502 mph). This is shown in Figure 3-5.

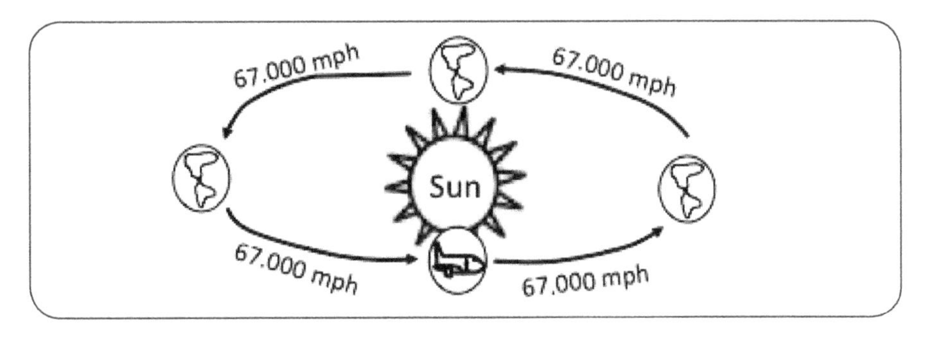

Figure 3-5. Speed depends on the object chosen as reference. *Movement of the person on the plane relative to the Sun is about 67,502 miles per hour (i.e., the Earth's movement plus that of the airplane and passenger, ignoring for the sake of simplicity, the rotational speed of the Earth's surface).*

If you then consider the Sun's movement around the galaxy you would get another relative speed, this time in comparison to the center of the galaxy as the frame-of-reference. This is shown in Figure 3-6. You could also measure the passenger and airplane's movement relative to other airplanes or you could measure the passenger's movement relative to other passengers. That is, any object can be selected as the frame-of-reference for measuring motion.

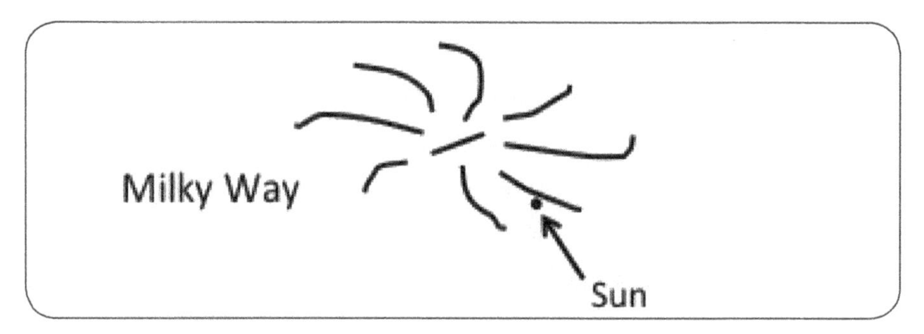

Figure 3-6. Our movements against some references are huge. *Movement of the Sun relative to the center of the Milky Way Galaxy is about 514,000 miles per hour.*

The point is that there is no common universal frame-of-reference for measuring movement. The movement of any object can only be sensed and measured by comparing it to the relative positions of other objects. But since there are an infinite number of objects that can be used as a relative reference for measuring motion, the key to measuring motion is in deciding what reference to use. That is, the first step in measuring motion is in specifying what frame-of-reference to use.

As mentioned earlier, the idea that all motion is relative to selected frames-of-reference was proposed well before Einstein. What Einstein did was combine this understanding that all motion is relative with the latest evidence about the invariance of the speed of light to develop a new and revolutionary understanding of the Universe, as we will see.

Because we live on Earth we are accustomed to measuring and describing our movements using Earth as our frame-of-reference. We often don't notice that other frames-of-reference are also available. As mentioned, the Sun, the moon, our galaxy, other galaxies, and even other objects on Earth, are all possible frames-of-reference that can be used for measuring our movements. Figure 3-7 summarizes the relationship among some of these.

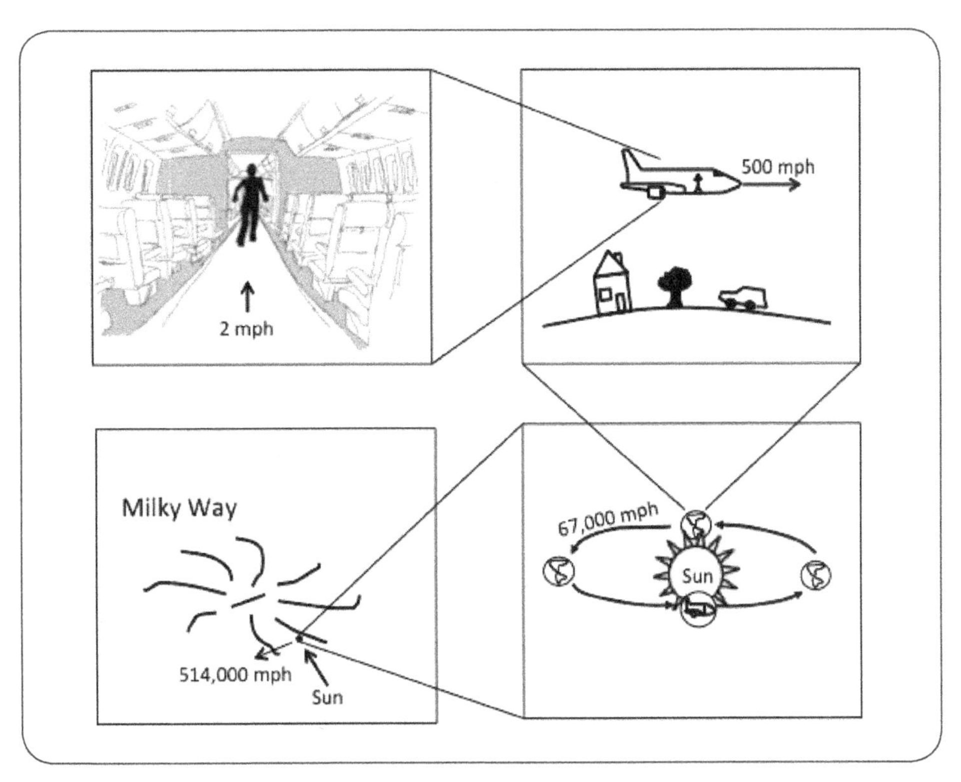

Figure 3-7. Many frames-of-reference are available for measuring motion. *Because we live on Earth we often forget that there are other possible frames-of-reference available.*

3.3 Incompatibility between relativity and the aether. Einstein's theories of relativity build on Galilean relativity and were incompatible with the existence of an "aether." This disagreement put Einstein at odds with many of his colleagues. The eventual acceptance of Einstein's theories of relativity ultimately led to the abandonment of belief in an aether.

There is no common frame-of-reference in the Universe that can serve as the standard for measuring the movement of objects, according to both Galileo and Einstein. That is, space is not defined by an aether or any other defining spatial structure against which all movement and locations can be judged. Instead, as noted, all movement and locations of objects are defined relative to the location and movement of other objects. This is one reason Einstein's theories are called theories of relativity — because they are based on the fact that the motion of any object is relative and only determined in comparison to (relative to) other objects, and that this has profound implications on the measurement of time and space, as we will see later.

Einstein's theories of relativity put him at odds with many of his colleagues who still believed in the existence of an aether. When the predictions of Special Relativity and later General Relativity were eventually confirmed and widely accepted, the idea that an aether existed was finally discarded.

3.4 The velocity of motion is relative too. The understanding that the motion of any object can only be measured relative to the location and motion of other objects led Einstein to realize that the relative speed of objects (i.e., their time-distance relationships) also depends on which object (i.e., frame-of-reference) is selected as the reference for measuring the motion. This formed the foundation for the Special Theory of Relativity.

Einstein used the understanding that all motion is relative (i.e., Galilean Relativity) as a foundation for defining the principles of Special Relativity. He asked us to envision a person on a train dropping a coin to the floor of the train car and then to consider the distance and speed that the coin traveled. Assume, for this example, the coin travels 3 feet and takes 1 second to travel this distance.[2] This means the coin moves at an average speed of 3 feet per second. This is shown in Figure 3-8.

Figure 3-8. We can measure the speed of motion relative to the train car. *The distance a coin travels when dropped to the floor of a train car is measured to be 3 feet. If the coin takes 1 second to land on the floor, we can say that the coin is traveling an average of 3 feet per second.*

2 The actual time it takes a coin to reach the floor from 3 feet would be less than 1 second. One second is used in this example just to illustrate the principle.

Now if a person on the ground was watching the train car moving past and views this same event of the coin being dropped, a different estimate of distance and speed is obtained. This time the coin appears to travel a longer distance, but still taking 1 second. In this example, the coin travels about 4 feet on the diagonal and therefore moves at an average speed of 4 feet per second. The speed and distance differences are the result of adding the movement of the train car to the movement of the falling coin. This is shown in Figure 3-9. Since both frames-of-reference, that of the train car and that of the ground, are equally valid, both results are equally correct. It is just that they are determined from two different frames-of-reference. In the first case, the coin's movement is measured relative to the train car. In the second case, the coin's movement is measured relative to the Earth and thus adds the movement of the train car (about 2.65 feet in this illustrative example) across the Earth to the movement of the coin within the train car (3 feet). This results in a longer distance traveled and a greater apparent speed. The coin moved 3 feet per second or 4 feet per second, depending on your point of reference or the selected frame-of-reference. Both are true.

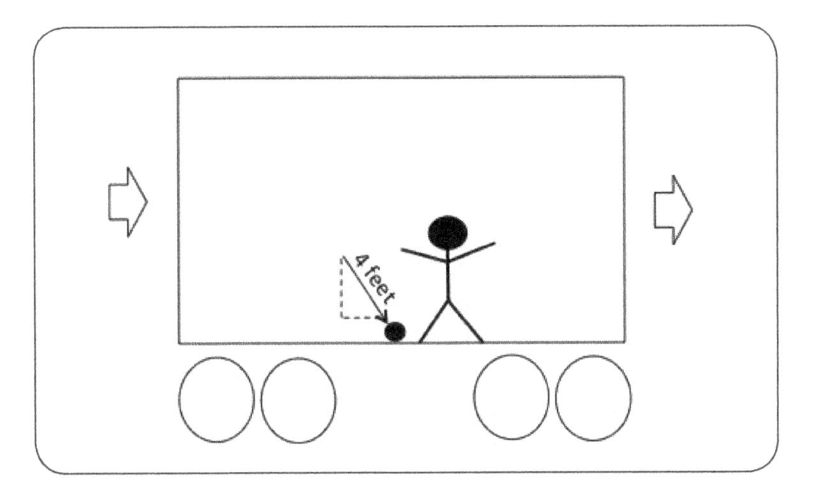

Figure 3-9. We can calculate the speed of motion relative to the Earth. *The distance a coin travels when dropped to the floor of a moving train car measured from the side of the tracks is 4 feet (the diagonal of a triangle that includes the vertical distance on the train car and the horizontal distance of the train car moving across the Earth). If the coin takes 1 second to land on the floor, we can say that the coin is traveling an average of 4 feet per second relative to the Earth.*[3]

3 Note that this hypothetical example is used for illustrative purposes. It ignores the fact that the coin would take a slightly curved path as the coin's movement would accelerate as it is pulled to Earth by Earth's gravity.

But what if the coin was replaced with the motion of photons in a beam of light? This is a special situation because we know that light always travel at speed "c" (186,282 miles per second). In this case, like for the coin, the photons appear to travel a greater distance as seen by an observer on the ground (i.e., the diagonal considering the vertical distance traveled within the train car and the horizontal distance traveled by the train car itself). However, in the case of a beam of light, the speed will not increase reflecting the longer distance. The speed of light is always 186,282 miles per second, or speed "c." So how can this be, since the same photons appeared to travel further? This is the question Einstein asked that led him to develop the first of his theories of relativity, the Special Theory of Relativity, which is explored in the next chapter.

As we will see in the next chapter, the explanation is that when objects are in relative uniform motion (at speeds approaching the speed of light) it is time itself that moves more slowly and space itself shortens accordingly.

Chapter 4
The Special Theory of Relativity

This chapter addresses the following topics and questions:

4.1 How Special Relativity works. *How does the strange behavior of light with its unvarying speed, together with Galilean Relativity, combine to form the basis of Einstein's Special Relativity?*

4.2 Why Special Relativity is called "special." *What is so special about it?*

4.3 Observer perspective is the key. *How does the perspective of the observer determine the effect of Special Relativity on time slowing and space compression? What is really moving and compressing?*

As noted at the end of the last chapter, Einstein developed his Special Theory of Relativity by combining: (1) the understanding that all motion is relative and can only be measured in relation to other objects (Galilean Relativity); (2) the fact that there is no universal frame-of-reference in space against which all motion can be compared; and (3) the knowledge that the speed of light is the same no matter how you are moving relative to the source of the light. This chapter introduces and describes Einstein's Special Theory of Relativity.

4.1 How Special Relativity Works. Galilean Relativity plus the behavior of light combine to form the basis of Einstein's Special Theory of Relativity. This section describes how it works.

What Einstein realized in developing the Special Theory of Relativity was that if all motion is relative and the speed of light does not change even when moving toward or away from the light source, it must be time itself that is going faster or slower. How else can the speed of light remain unchanged? This chapter examines and illustrates this.

The effects of Special Relativity can be difficult to appreciate and understand because they only become evident when traveling at speeds approaching the speed of light, speeds not experienced in our day-to-day lives. Even speeds experienced on the fastest supersonic aircraft don't come close to that needed for making the effects of Special Relativity observable. Therefore to appreciate the concepts of Special Relativity and describe their effects, it helps to imagine the conditions under which they occur. This chapter does this. It presents imaginary situations under which Special Relativity effects are evident.

The imaginary situation again considers the movement of a train car, but a train that is capable of moving at near the speed of light where the effects of Special Relativity become significant.[1] The train car has a light bulb on the ceiling and a mirror on the floor directly below it. Now also imagine that the light bulb emits a single photon of light that travels from the light bulb

1 The principles of relativity are applicable at all speeds, even the relatively slow speeds experienced on trains. However, the effects of relativity at such low speeds are extremely small, in fact so small that they are not measurable or noticeable. It is only at speeds approaching the speed of light that the effects of Special Relativity are significant. This is why, for the case of dropping a coin on a train car, the measured speeds of the coin relative to the car can be added to the speed of the train; because the effects of relativity are so small at such slow speeds that they can be ignored.

to the mirror on the floor and back to the light bulb. Also, finally, imagine that a scientist has a stopwatch capable of measuring very short amounts of time and uses it to measure how long it takes for the photon to make the trip. This allows the scientist to determine how fast the photon is moving.[2] This is shown in Figure 4-1.

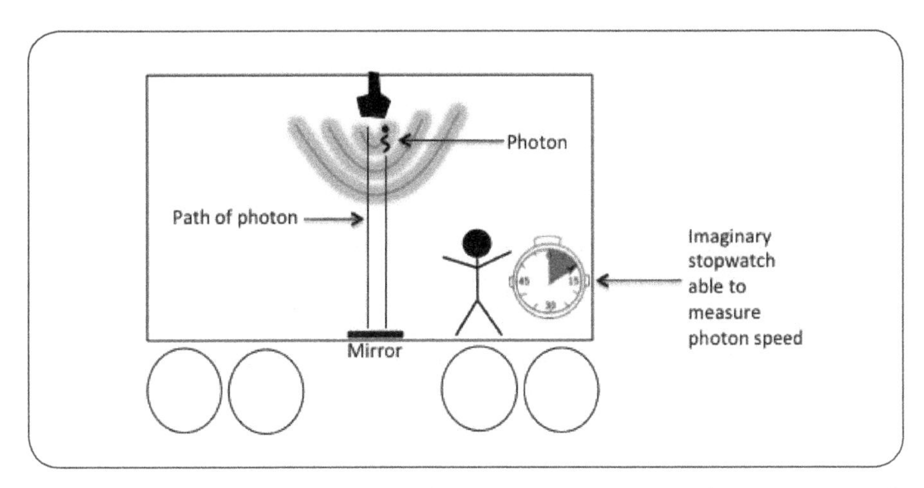

Figure 4-1. The speed of light is measured within a train car. *This is done using a special (imaginary) stopwatch capable of measuring extremely short periods of time. The scientist measures the speed of the photon as it goes from the light bulb on the ceiling to the mirror on the floor and back to the ceiling. As expected, the photon is found to be traveling at speed "c."*

Using the hypothetical stopwatch, the scientist on the train determines the speed of the photon. This is done by measuring the distance the photon travels (from the light bulb to a mirror on the floor and back to the light bulb) and the time taken by the photon to cover this distance. The speed of the photon is determined by simply dividing the distance traveled by the time taken, just like we measure the speed of a car in miles per hour (miles divided by hours). As expected, and in accordance with prior measurement of light speed described earlier, the light on the train is found to travel at 186,282 miles per second, that is, speed "c."

Now imagine a second scientist on the ground also measures the time it takes for the photon to travel from the light bulb to the mirror and back to the light bulb. For this scientist the distance traveled seems further because of the movement of the train car, which in this example is moving at near the

2 This situation represents one of Einstein's "thought experiments" that he used to illustrate the workings of relativity at a time when the needed scientific experimentation could not be done using available methods and instrumentation. Thought experiments are used throughout this book based on those used by Einstein. However, the underlying workings of relativity have since been experimentally verified. Some of these verifications are summarized in Chapter 11.

speed of light. So when the photon's speed is calculated you might expect a faster speed since a greater distance is covered. But unlike for the case of the dropped coin described earlier, the scientist on the ground finds that the speed of the photon is speed "c," the same speed as the scientist on the train found. As noted earlier, this is what scientists find when they measure the speed of light regardless of any relative motion of the light source; light (and all forms of electromagnetic waves such as radio waves) always moves at speed "c." This is consistent with many experiments measuring the speed of light from different light sources in relative motion (e.g., like when scientists measure the speed of light coming from distant stars as Earth moves toward and away from them in its orbit around the Sun. The speed is always the same, 186,282 miles per second!) And this is what would be found in this hypothetical situation. This is shown in Figure 4-2.

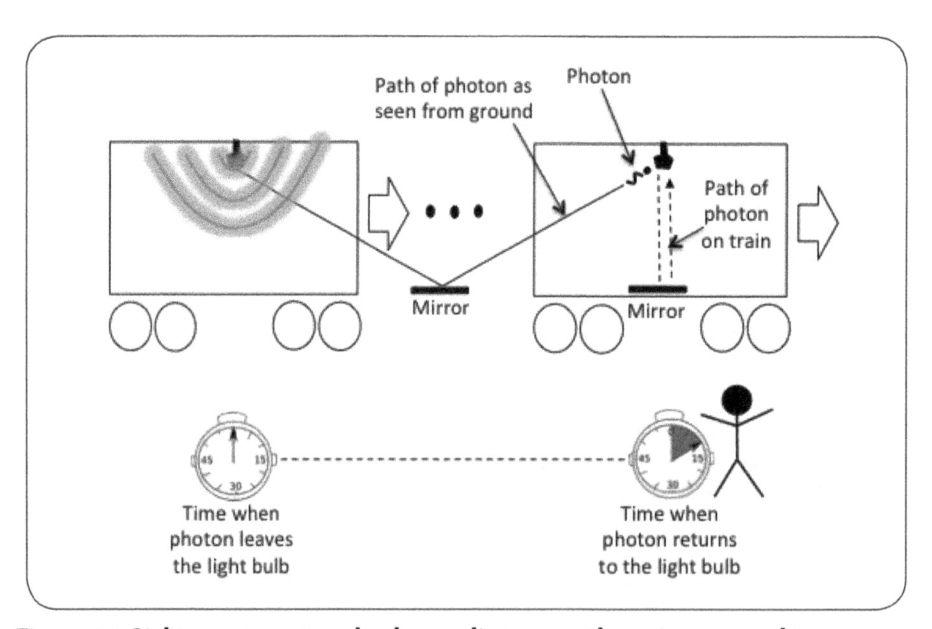

Figure 4-2. Light appears to travel a shorter distance on the train compared to as seen from the ground. *A scientist on the ground measures the speed of the photon on the train. The distance traveled by the photon appears to be further as seen from the ground while the distance it travels on the train is less. But we know that the speed of the photon is always the same (186,282 miles per second or speed "c").*

But how can this be? The scientist on the ground sees the photon as traveling further. How can the speed of the photon be the same for both the scientist on the train and the scientist on the ground? How can both scientists measure the same speed for the photon when in one case a greater distance is covered?

The immediate answer to this riddle is: that it just is. This is what scientists find when they take the measurements. Just like when scientists measured the speed of light coming from a star, both when we were approaching the star and when we were moving away from it, the speed of light (and of all electromagnetic waves) is the same regardless of perspective of the observer – even though it doesn't match our intuition of how light ***should*** behave. It is how light ***does*** behave. Whether we understand the reasons for these surprising observations, or agree that it should be this way, doesn't change the fact that this is what we find. So how do we explain this?

The answer to this puzzle, and the brilliant idea offered by Einstein, is that time itself moves more slowly on the moving train, from the perspective of the scientist on the ground. Einstein came to this conclusion by reasoning that if: (1) the photon within the moving train frame-of-reference appears to travel a shorter distance; and (2) the speed of the photon is the same speed "c" (186,282 miles per second), regardless of the frame-of-reference (the moving train or stationary Earth); then (3) time itself must move more slowly on the moving train compared to time as experienced by the scientist on the ground. This is shown in Figure 4-3.

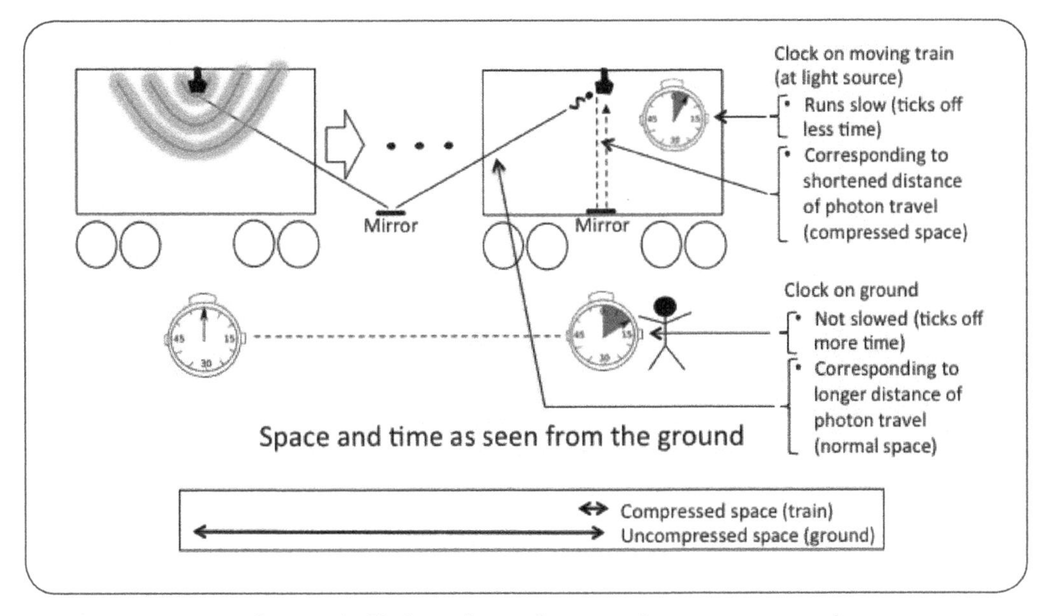

Figure 4-3. Since the speed of light is always the same, then time must go slower on moving platforms. *Because the speed of light is always speed "c," a scientist on the ground sees time on the train as going slow. That is, clocks on the train will tick off less time corresponding to the apparently shortened distance (compressed space) traveled by photons in that moving frame-of-reference.*

Yes, it's true! The clock on the moving train, as observed by the scientist on the ground, ticks off less time during the event of the photon going from the light bulb to the mirror and back to the light bulb. This makes sense because the distance the photon traveled within the moving train was less than the distance as seen from the ground. Since the speed of the photon is always speed "c," time itself on the moving train must go more slowly, corresponding to the shorter distance travelled by the photon on the train. How else could the photon speed be the same? That is, the clock on the train as observed from the ground ticks off less time compared to the clock on the ground. This corresponds to to the shortened photon travel distance (compressed space) within the moving train frame-of-reference! The shorthand for remembering the relationship between moving platforms and the slowing of time is that *"moving clocks run slow."* Scientists use this phrase to help remember that it is on moving platforms that time is running slower. It will come in handy for you to remember this effect as you read the rest of this book.

According to Einstein, this is how space and time are related and how the Universe works. Space and time are not separate things. They are interrelated and subject to the effects of relative movements — hence the theory of relativity! And these effects become meaningful and measurable at relative speeds approaching the speed of light. The distance a photon travels in a moving frame-of-reference is less (i.e., space is compressed) and this corresponds to clocks running more slowly (i.e., ticking off less time). Measurements of both space and time vary when measured from different frames-of-reference that are moving relative to each other. Chapters 9 and 10 will show how these concepts play out in the workings of the Universe.

Finally, it must be noted (again) that these time and space differences are only meaningful (and measurable) for relative speed differences near the speed of light (much, much faster than trains go). I only use the train example, like Einstein did, as a way of illustrating the principle, even though the space and time effects for normal train speeds are extremely small and not noticeable or measurable. Nevertheless, if trains could go at speeds close to the speed of light, the effects would be as described. In fact, these effects have been demonstrated experimentally for objects that do move at speeds near that of light, and have even been shown to occur on fast moving aircraft. The experimental evidence for this is summarized in Chapter 11.

4.2 Why Special Relativity is called "special." Special Relativity is "special" because it only applies in certain limited situations. Calling it "special" recognizes the limited (special) nature of the situations under which it applies and the fact that it doesn't apply more generally.

Special Relativity only applies to objects that are in "uniform motion." This means that it applies only to relative motion that is not changing speed and not changing direction (i.e., accelerating). As we will see later, this is also called "inertial motion" because without any acceleration, an object's movement is determined by its inertia. When accelerations are present, the laws defined by Special Relativity do not apply and can't be used to calculate time and space variations.

Under conditions of acceleration, a more robust, more complete theory of relativity is needed. This more complete theory is called the General Theory of Relativity. Einstein developed the more complete, and mathematically more difficult, General Theory of Relativity well after he completed the more limited Special Theory of Relativity. The General Theory of Relativity is covered in Chapter 8.

4.3 Observer perspective is the key. The perspective of observers is central in defining the effects of Special Relativity.

Now since all motion is relative, it is just as possible to consider the opposite perspective for the train example with the train as the stationary frame-of-reference and the Earth whizzing past. To illustrate this, we will again consider a single photon observed from both frames-of-reference. This time, the photon is emitted from a light bulb on a pole on the ground. A mirror on the ground reflects the photon back to the light bulb (see Figure 4-4). A scientist on the train observes this event (see Figure 4-5). This time, the scientist on the train sees the photon as experienced within the moving ground-based frame-of-reference traveling a shorter distance and therefore perceives clocks on the ground to be running slower (ticking off less time), again with the speed of the light always at speed "c."

The point is that the slowing of time and commensurate compression of space works both ways! There is no preferred frame-of-reference for measuring time and space; all frames-of-reference are equally valid. The scientist on the train sees the pole moving past and the clocks on the ground running slow. The scientist on the ground sees the train moving past and clocks on the train running slow. And both are right!

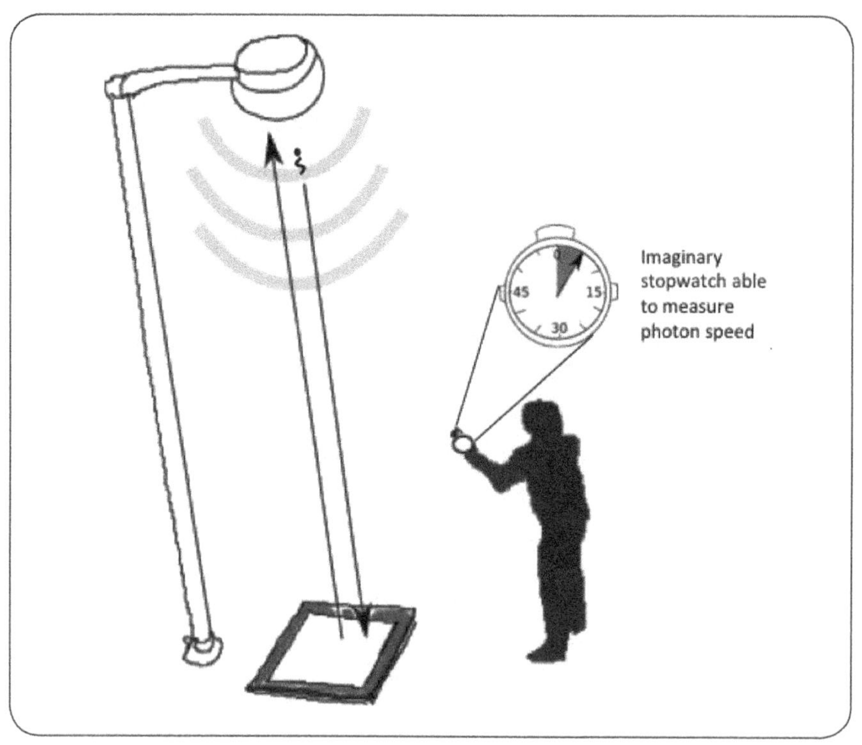

Figure 4-4. The speed of light is measured on the ground. *A single photon is emitted by a light bulb on a pole and reflected back to the light bulb by a mirror on the ground. The photon travels at speed "c."*

And this makes perfect sense given that the speed of light is fixed. To the scientist on the train, the ground appears to be moving past and the photons within the ground-based frame-of-reference appear to travel a shorter distance. So therefore, the clocks on the ground appear to run slow (tick off less time) using the same logic and analysis as for the opposite case described above. It works relative to the perspective of the observer.

Again, the shorthand that scientists use for remembering the effects of relative motion is that *"moving clocks run slow."* That is, the clocks in the frame-of-reference that are in apparent motion will be perceived to run slow (tick off less time) compared to clocks in the same frame-of-reference as the observer. This is shown in Figure 4-5. Illustrations presented in Chapter 9 will explicitly show how space shortens as time slows. It will be obvious when you see it.

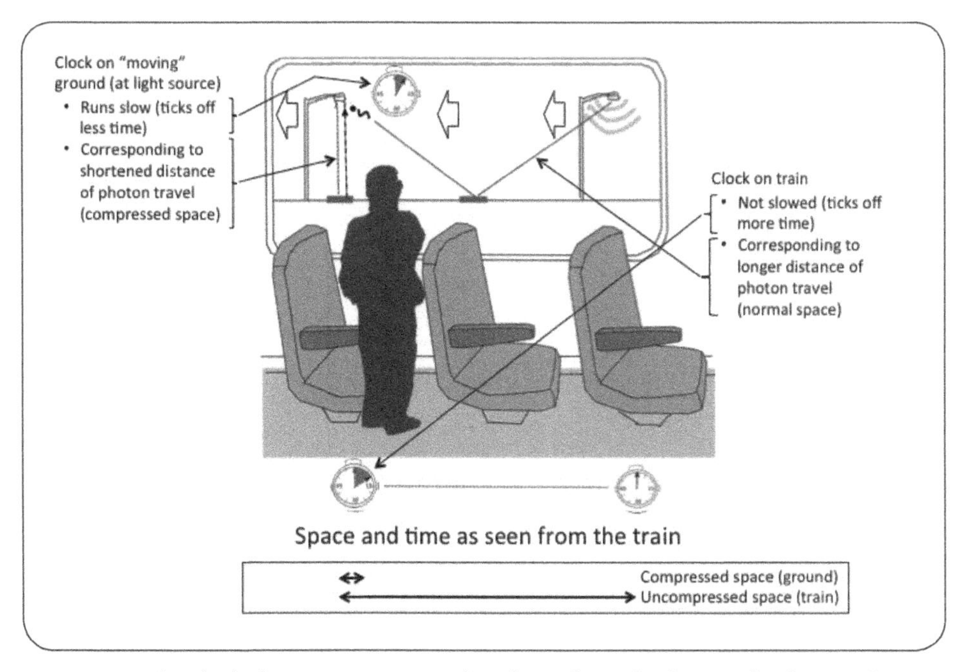

Figure 4-5. The clock that appears to run slow depends on the frame-of-reference from which measurements are taken. *A scientist on the train observes light emitted from the light bulb on the pole outside the train. From this train-bound perspective, these photons appear to travel a shorter distance as experienced by those on the ground. Therefore, from the perspective of the moving train, clocks on the ground appear to run slower (ticking off less time).*

Einstein mathematically worked out the relationships between relative speed and the degree of corresponding time slowing and space compression. It is not necessary to understand how these mathematical relationships are derived. But, for those interested and who have just a high school level understanding of geometry, the derivation of these relationships is not difficult. They can be understood using the Pythagorean theorem (that's the theorem that says the sum of the two short sides of a right triangle, each squared, equals the square of the longest side, the "hypotenuse," ($c^2 = a^2 + b^2$). This is presented in Appendix A. You will be surprised at how easy it is to understand!

Chapter 5
Clocks

This chapter addresses the following topics and questions:

5.1 Clocks. *What is a clock and how are clocks related to time?*

5.2 Light-clocks. *How can hypothetical clocks that use the movement of light to measure time provide a deeper understanding of relativity?*

5.3 Clocks and time. *Do clocks accurately measure time? Will regular clocks agree with hypothetical light clocks?*

Time and its measurement are central for understanding the concepts of relativity. Because clocks are how we measure time, and therefore the effects of relativity, this chapter is dedicated to the discussion of clocks. It describes how clocks work and how they provide a measure of time.

5.1 Clocks. Clocks are central to understanding relativity. This section answers the question, what is a clock and how are clocks related to time and relativity?

There are many kinds of clocks in use today. Various forms of clocks have been used throughout history. All have one thing in common. They all involve an object that repeats the same motion over and over with an unchanging time interval between repetitions. By counting the repetitions, or measuring progress within a repetition, it is possible to quantify or estimate the amount of time that has passed. Early clocks, such as sundials, relied on astronomical objects, such as the Sun or Moon. The number of times the Sun appears to go around the Earth marked the passage of days, while the extent of movement of the Sun within a single day allowed the estimation of the passage of hours.[1]

More modern clocks are mechanized and have three parts: (1) a power source, (2) an oscillator (the object that engages in repeated motions), and (3) the display, which translate the oscillator movements into a display of time passed. For example, in a grandfather clock, weights provide the power source that keeps the pendulum (the oscillator) moving and driving the hands of the clock (the clock display). In mechanical watches, a spring provides the power source, while in electronic watches batteries provide the power. The oscillator in modern clocks often use known vibration characteristics of specific molecules as the oscillator. For example, many clocks today use known vibration characteristics of quartz crystals to count the passage of time. Some early electronic clocks used the vibrations of a tuning fork as the oscillator. Figure 5-1 shows an example of clock technology highlighting major components.

1 The repetition involving the sun that provides a measure of the passage of each day is, of course, actually the repetitive movement of the Earth's rotation about its axis. Since this is reflected in the appearance of the sun moving around the Earth, and since in ancient times this is what was believed to occur, the movement associated with the passage of each day is described as the movement of the Sun around the Earth.

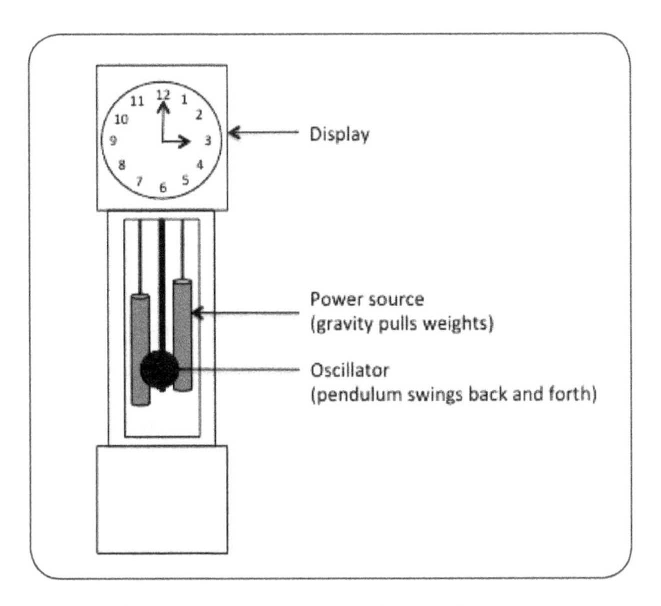

Figure 5-1. Clocks have three main components: the display, power source and oscillator. *Here is a grandfather clock showing its major components.*

5.2 Light-clocks. Hypothetical clocks that use the movement of light as a measure of time provide significant insight into how relativity works. In light-clocks, the movement of light inside the clock serves as the oscillator. Light-clocks are introduced here and used throughout the rest of the book to help make relativity understandable. Light-clocks are only hypothetical – they are used in this book only to illustrate the principles of relativity.

The earlier discussion of Special Relativity associated time with the movement and behavior of light on a moving train. It showed that since we know that light always moves at 186,282 miles per second its movement is related to time and distance (space) in a fixed, defined way. A light clock uses this knowledge by counting the number of times a photon can move between two points in the light clock that are separated by a known distance. By counting the number of times light can travel between them, we know the amount of time passage based on the speed of light. This approach to measuring time forms the basis for light-clocks. While light-clocks are hypothetical and don't actually exist on Earth, they are used here, as in other books on relativity, to illustrate the principle of how light can be used to measure time. If one were able to build a light-clock it would work as follows.

In a light-clock, photons move back and forth between a light source and a mirror located at opposite ends of a tube. The light source powers the light-clock; the photons move to the mirror and reflect back to the light source triggering the next light pulse (the next photon). The back and forth movement of the light (the photons) serves as the clock's oscillator. The number of pulses of light that are able to travel back and forth in the tube measures the amount of time that passes. Each two-way transit of the light in the tube represents one tick of the light-clock. Figure 5-2 provides a graphic illustration of a light-clock.

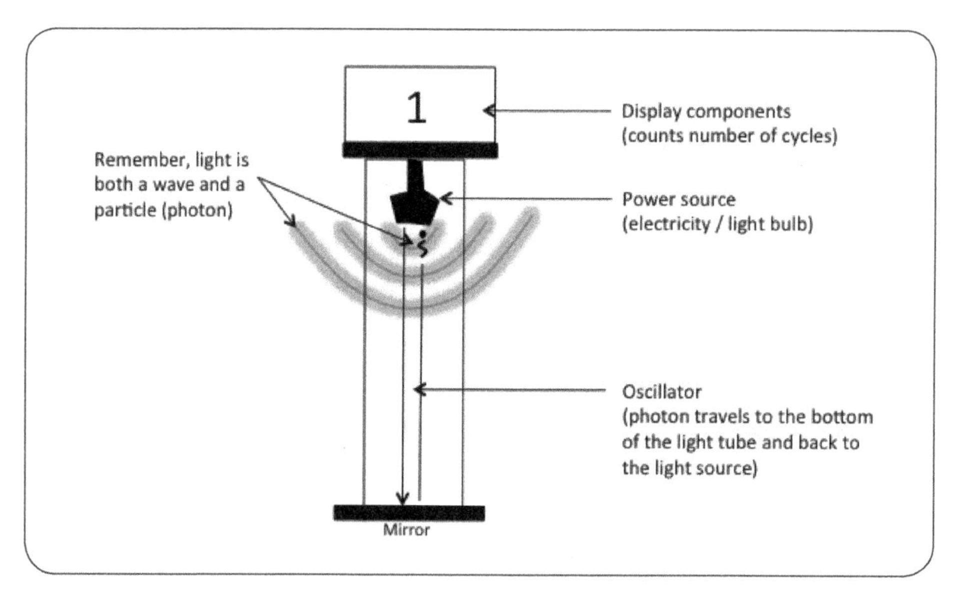

Figure 5-2. How a light-clock works. *The hypothetical light-clock uses the movement of photons to measure time and the effects of relativity.*

The previous discussion of the transit of a photon from ceiling to floor and back to the ceiling on the train is really just an application of a light-clock (without the enclosing box or "tube"). Each movement of a photon to the bottom of the train car and back to the light source is equivalent to one tick of a light-clock. In the train illustration we saw that because of Special Relativity there is an interaction between time and the relative motion of the train when speeds approaching "c" are considered. That is, a light-clock on a train runs slower when viewed from a frame-of-reference against which the train is in relative motion, such as for an observer on the ground.

The perception of slowed time corresponds to the unchanging speed of light together with the shorter distance the light travels on the "moving" platform compared to the "stationary" platform. "Light-clocks" are used to illustrate the amount of time slowing and therefore to describe the effects of relativity.

Since the effects of relativity are only evident and measurable at relative speeds approaching that of light, the use of a clock where the oscillator itself moves at the speed of light is perfect for the task. We'll use a hypothetical light-clock for our discussion of relativity from here forward. Figure 5-3 illustrates the use of a light-clock for measuring the effects of Special Relativity on a rocket ship moving relative to the Earth. It illustrates the effect of relative motion on time by showing a light-clock observed across two frames-of-reference: (1) the "moving" rocket ship; and (2) the "stationary" frame-of-reference of the Earth. Of course, we know that either frame-of-reference can be considered to be moving. In this example, we have chosen to consider the rocket ship to be the moving platform. As in the train example, the light appears to travel a longer distance when viewed from the "stationary" Earth and a shorter distance as experienced on the "moving" rocket ship. With the constant speed of light, time is slowed on the rocket ship (moving) frame-of-reference as seen from the (stationary) Earth frame-of-reference. Remember, *"moving clocks run slow."* Note that these figures provide a simplistic two-dimensional view of time slowing and space compression that really happens in a four-dimensional realm. Two-dimensional pictures are used here, as Einstein did, to make it easy to see the effects.

The stopwatch icons and space compression graphic (ruler) in Figure 5-3 will be used throughout this book to represent effects of relativity on time and space. The stopwatches associated with the two frames-of-reference represent the comparative slowing of time on the "moving" platform. The shortened ruler is used to symbolize the space compression associated with the shorter light path on the moving platform. Space is always compressed along the direction of travel as symbolized by the shortened ruler.

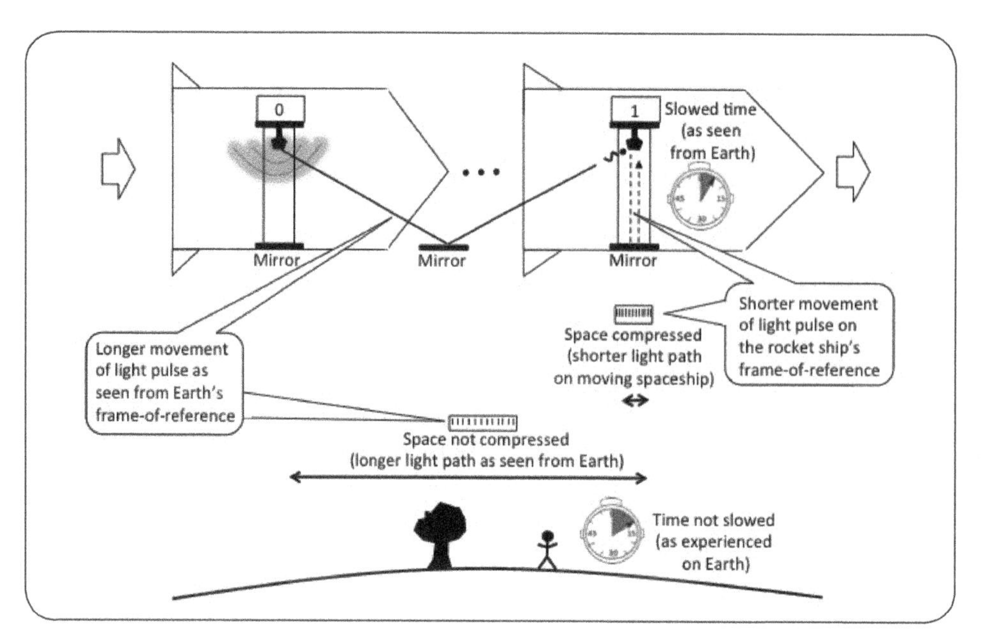

Figure 5-3. A light-clock can illustrate time slowing on a moving object. *This figure shows the use of a light-clock to illustrate the Special Relativity effects of a rocket ship moving relative to Earth. The stopwatch and space compression graphics are used to remind us that time is slowed and space compressed (shorter light path) on moving platforms.*

5.3 Clocks and time. Can we say that the clocks we use on Earth accurately measure time? Would they agree with hypothetical light-clocks? This section answers these questions.

Hypothetical light-clocks are unique because their oscillators work at the speed of light, a speed that for light is fixed and unvarying. But do light-clocks agree with "real" clocks that we use on earth and that we have come to trust, such as atomic clocks, quartz crystal clocks and Timex™ clocks? And do these "real" clocks actually provide an accurate measure of time? Intuitively, the answer is yes, assuming that the pendulum swings of a grandfather clock, for example, are at regular intervals and each takes the same amount of time. And similarly this would be true for crystal vibrations and vibrations of the atoms of atomic clocks, as long as each vibration is regular and has fixed duration relative to actual time. These real clocks should provide a consistent and accurate measure of the passing of time. But can we say this with certainty?

What we can say from experimental evidence is that yes, "real" clocks do agree with hypothetical light-clocks when considering the predictions of relativity. We can say this because when scientists have used real atomic clocks to determine whether relative motion actually is reflected in a slowing of time, as hypothetical light-clocks and Special Relativity would predict, they have found that the real clocks do run slower. The results have agreed with the predictions. For example, when atomic clocks were put on supersonic airplanes scientists found that the moving atomic clocks tick off less time compared to identical atomic clocks that stayed on Earth. These experiments agree exactly with the expected results. And to assess whether these atomic clocks are related the underlying construct of time, scientists have compared time as measured by real world processes such as the decay of radioactive particles. They have found that both the atomic clocks and the rate of radioactive decay slow by the same amount, and in accordance with the predictions of Special Relativity.

These and other experiments that demonstrate the validity of Einstein's theories of relativity are summarized in Chapter 11. So, yes, hypothetical light-clocks do agree with "real" clocks; both are sensitive to the effects of Special Relativity. I will continue to apply the hypothetical light-clocks to graphically illustrate the effects of relativity in subsequent chapters.

Chapter 6
Time and Space

This chapter addresses the following topics and questions:

6.1 Time slowing. *How much does time slow as a function of increased relative motion?*

6.2 Space compression. *How much space compression occurs as a function of increased relative motion?*

6.3 Bi-directional application. *Can the slowing of time and space compression really work both ways with either frame-of-reference selected as the observational ("stationary") reference frame?*

Einstein's Special Theory of Relativity not only showed us why time slows and space shortens (compresses) when relative motion approaching the speed of light is experienced from the perspective of an observer; it also mathematically specified the relationship between relative motion and the associated slowing of time and compression of space. His formulas determine how much time slows and how much space compresses based on the relative speeds involved.

This chapter describes these quantitative relationships to build a deeper appreciation of the Special Theory of Relativity. It first reviews the relationships and then presents the main mathematical formulas for those interested. (It is not necessary that every reader understand the formulas. They are included only for those who want to see them and gain a deeper appreciation of relativity and its underlying principles.)

Again, remember there is no motion except relative motion, that is, motion between objects (i.e., frames-of-reference). How else can you tell that something is moving? Special Relativity affects time and space as judged between objects that are moving relative to each other.

6.1 Time slowing. This section describes the quantitative relationship between relative motion and time. It addresses the question: how much does time slow as a function of relative motion?

As described in the previous chapters the slowing of time on a moving platform occurs because of Special Relativity. It derives from the fact that the speed of light remains constant even when the light source is moving. The light appears to travel a shorter distance within the "moving" platform as seen from the "stationary" platform due to the relative motion. With the speed of the light fixed, time must be slower on the moving platform where the light path is shorter (i.e., space is compressed).

In addition to figuring out why time slows Einstein determined the amount of slowing based on the relative speeds involved. He developed a correction factor called the "relativistic correction factor." The relativistic correction factor is used to determine the amount of time slowing that occurs based on the relative speeds of the two platforms. At speed "c" there is complete time slowing (time stops) and complete space compression. Much more on this later. Here's the relativistic correction formula for those interested.

OPTIONAL: For those interested the formula for calculating the time correction is shown below.

Einstein's Relativistic Correction Factor for Time Slowing

The formula for calculating the amount of time slowing based on velocity and the fixed speed of light is:

$$t_{(moving)} = t_{(stationary)}\sqrt{1-\frac{v^2}{c^2}}$$

Where:

$t_{(moving)}$ = time passage on the moving platform (as seen from the stationary platform)
$t_{(stationary)}$ = time passage on the stationary platform that serves as the reference for judging the motion
v = velocity of the moving platform relative to the stationary platform
c = the speed of light (1 light-year per year)

For those who want to see why this formula is correct a very simple description of the formula's derivation using the Pythagorean theorem (from high school geometry) is included in Appendix A. (Take a look. You'll be surprised how easy it is to understand!)

So let's see how Einstein's relativistic correction factor works using an example. We'll consider the example shown in Figure 6-1. In this example a rocket ship is flying past Earth on its way to a nearby star. The star is 10 light-years away from Earth. In this example we assume that the star is not moving with respect to the Earth. The rocket ship is moving at 80% of the speed of light between the two (relative to the Earth / star frame-of-reference).[1]

1 This example was taken from Richard Wolfson's book *Simple Einstein: Relativity Demystified.*

We can easily calculate how long it will take the rocket ship to get to the star measured from the Earth / star frame-of-reference ($t_{(stationary)}$ the lower stopwatch). We can also calculate how much time passes as experienced on the rocket ship frame-of-reference ($t_{(moving)}$ upper stopwatch) as seen from the "stationary" Earth / star. We'll use Einstein's relativistic correction factor to see how much time passes for those on the rocket ship compared to those on Earth as judged from Earth.

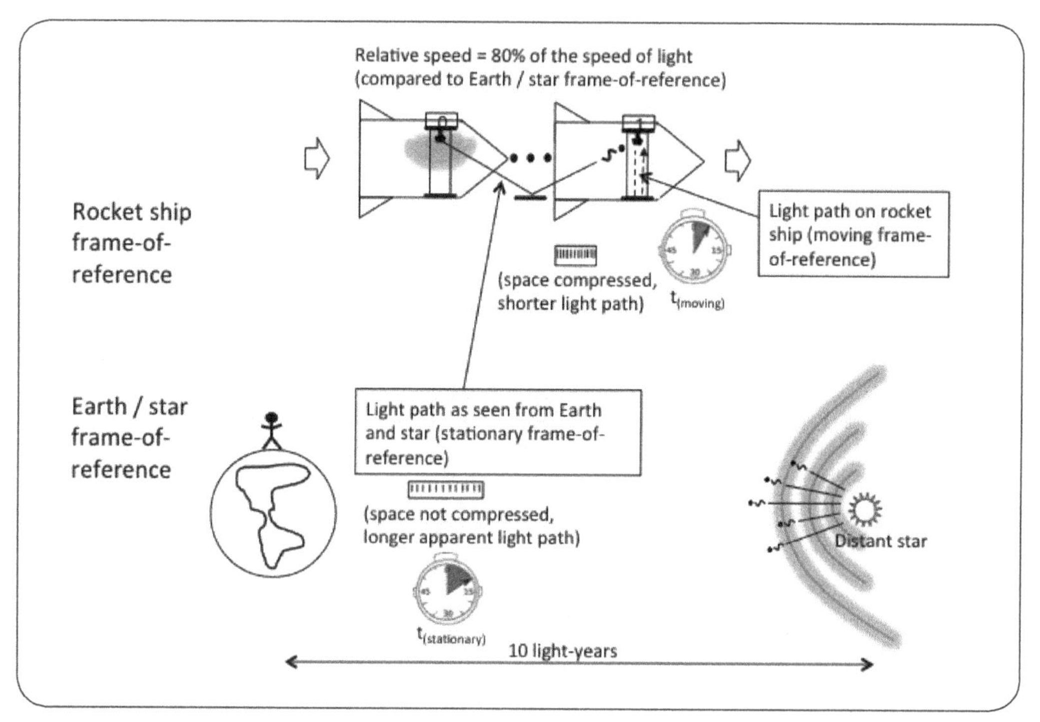

Figure 6-1. How high-speed travel affects the passage of time for the "moving" rocket ship. *In this illustration the rocket ship is moving at 80% of the speed of light relative to the Earth and star. Because the Earth / star is the "stationary" frame-of-reference the elapsed time on the moving rocket ship is less as measured from the Earth.*

First how long does it take the rocket ship to travel from the Earth to the star as measured from the Earth / star frame-of-reference (the lower stopwatch)? This is easy just like figuring out how long it will take to drive somewhere when the distance and driving speed are known. For example to figure out how long it will take to drive 120 miles at 60 miles per hour you would divide 120 miles by 60 miles per hour to figure out that it would take 2 hours.

Here's the formula:

Distance = speed × time
120 miles = (60 miles/hour) × (2 hours)

Now let's do the calculation for the trip between the Earth and the star.

In this case the distance is 10 light-years and the speed is 80% of speed "c" so:

Distance = speed × time
10 light years = (0.8) × (c) × (time)
10 light years = (0.8) × (1 light year/year) × (time)

And figuring out how much time it will take by solving for "time":

$$\text{Time} = \frac{10 \text{ light years}}{(0.8)(1 \text{ light year/year})}$$

$$\text{Time} = 12.5 \text{ years}$$

This means it takes the rocket ship 12.5 years to go between the Earth and star when traveling at 80% of the speed of light as measured from Earth. This makes sense because if the rocket ship were traveling at the speed of light the trip would take 10 years since the star is 10 light-years away. But since the rocket ship is traveling slower than the speed of light (80% of the speed of light) the trip takes a little longer at 12.5 years (an extra 2.5 years).

Second how long does it take to travel from the Earth to the star on board the rocket ship (the upper stopwatch)? We know that the clock on the rocket ship is running slow *("moving clocks run slow")* compared to the clocks on the "stationary" Earth. We now apply the relativistic correction factor. In this example (for objects moving at 80% of the speed of light) it turns out that the correction factor is 60% (calculated below for those interested). This means that clocks on the moving rocket ship run 60% as fast as the "stationary" Earth and star clocks. When we apply this relativistic correction factor (60%) we find that the trip for those on the rocket ship only takes 7.5 years as shown below.

Time *(on the moving platform)*	=	Normal Time *(on "stationary" platform)*	×	Correction Factor
7.5 years	=	12.5 years	×	60%

OPTIONAL: For those interested this calculation is shown below:

Determining and Applying Einstein's Relativistic Correction Factor for Time Slowing

Here's the calculation for determining and applying Einstein's correction factor to find the amount of time slowing for the above example.

First, determine the correction factor:

$$t_{(moving)} = t_{(stationary)}\sqrt{1 - \frac{v^2}{c^2}}$$

$$t_{(moving)} = (12.5\ years)\sqrt{1 - \frac{(.8c)^2}{c^2}}$$

$$t_{(moving)} = (12.5\ years)\sqrt{1 - \frac{.64c^2}{c^2}}$$

$$t_{(moving)} = (12.5\ years)\sqrt{1 - .64} = \sqrt{.36} = .6 = 60\%$$

Next, apply the correction factor to determine the time it takes on the moving platform:

$$t_{(moving)} = t_{(Earth-star/reference)} \times (Einstein's\ correction\ factor)$$

$$t_{(moving)} = (12.5\ years) \times (Einstein's\ correction\ factor)$$

$$t_{(moving)} = (12.5\ years) \times (.6)$$

$$t_{(moving)} = 7.5\ years$$

This means that as seen from Earth, time on the rocket ship moves 60% as fast as time on Earth, the stationary frame-of-reference. Therefore for those on the rocket ship it takes 7.5 years to get to the star much less than the 12.5 years that pass on Earth clocks. This is consistent with what we now know; that is that clocks run slow (tick off less time) on moving platforms. Einstein offered an example of how this affects the aging of two twins moving relative to each other. Einstein's "twin paradox" is described in Chapter 9.

6.2 Space compression. This section describes the relationship between time slowing and space compression. It answers the question: why do both occur together and how much space compression occurs as a function of relative motion?

So how can the rocket ship cover the distance of 10 light-years in only 7.5 years since the star is 10 light-years from Earth? This would require going faster than the speed of light, which we know is not possible. The answer is that the space traveler actually experiences the journey as a much shorter distance. In addition to time being slowed, space is compressed making the distance covered less than 10 light-years for the moving rocket ship. Movement at speeds approaching the speed of light has the effect of slowing time ***and*** compressing space.

We can easily figure out how much less distance the occupants of the rocket ship experience by applying the simple formula for distance as we did earlier; but this time using the new number we just figured out for travel time, 7.5 years. When we do this we find the distance traveled is 6 light-years. This is shown below.

$$\text{Distance} = \text{speed} \times \text{time}$$
$$d_{(moving)} = \text{speed} \times t_{(moving)}$$
$$d_{(moving)} = (0.8\ c)\ (7.5\ \text{years})$$
$$d_{(moving)} = (0.8 \times 1\ \text{light year/year})(7.5\ \text{years})$$
$$d_{(moving)} = 6\ \text{light years}$$

Where:

$d_{(moving)}$ = distance covered by the moving platform (measured from the stationary platform)

$t_{(moving)}$ = time passage on the moving platform (measured from the stationary platform)

speed = velocity of the moving platform relative to the stationary platform

c = the speed of light (1 light-year per year)

So we see that less time passes during the trip on the rocket ship, the moving platform (slowed time: 7.5 years on the rocket ship compared to

12.5 years on the "stationary" Earth). This corresponds to less distance covered by the rocket ship, the moving platform (compressed space: 6 light years for the rocket ship compared to 10 light years as measured from the "stationary" Earth). You will notice that the amount of space compression is in the exact proportion as the amount of time slowing. That is, for the example described here, the amount of time slowing and space compression are both 60% as described below.

- Time for the trip on the rocket ship (moving platform) vs. the "stationary" Earth / star frame-of-reference is 7.5 years vs. 12.5 years (60% time slowing)
- Distance of travel on the rocket ship (moving platform) vs. the "stationary" Earth / star frame-of-reference is 6 light-years vs. 10 light-years (60% space compression)

This is not a coincidence. Einstein's Relativistic Correction Factor that we applied to calculate the amount of time slowing is the same as the correction factor for calculating space compression. It applies to both time slowing and space compression. And this makes sense; it takes less time to travel a shorter distance.

OPTIONAL: For those interested the formula for calculating the distance (space) correction factor is shown below.

Einstein's Relativistic Correction Factor Applied to Space

The formula defining the amount of space compression based on velocity is:

$$d_{(moving)} = d_{(stationary)} \sqrt{1 - \frac{v^2}{c^2}}$$

Where:

$d_{(moving)}$ = space on the moving platform
$d_{(stationary)}$ = space on the stationary platform
v = velocity of the moving platform relative to the stationary platform
c = the speed of light (1 light-year per year)

Again for those who want to understand why this formula is correct a very simple description of the formula's derivation using the Pythagorean theorem (from high school geometry) is included in Appendix A.

6.3 Bi-directional application. Time slowing and space compression work both ways. That is, either frame-of-reference can be selected as the observational ("stationary") frame.

As discussed earlier all motion is relative. It is possible to consider the rocket ship to be stationary and the Earth and star to be moving. From this alternative perspective time slowing and space compression are evident on the Earth / star frame-of-reference as seen by observers on the rocket ship. Since the same relative speeds are involved (and only the viewing perspective has changed) the same amount of time slowing and space compression is experienced. But this time the time slowing and space compression appears to happen on the Earth / star frame-of-reference as seen from the rocket ship. This is shown in Figure 6-2.

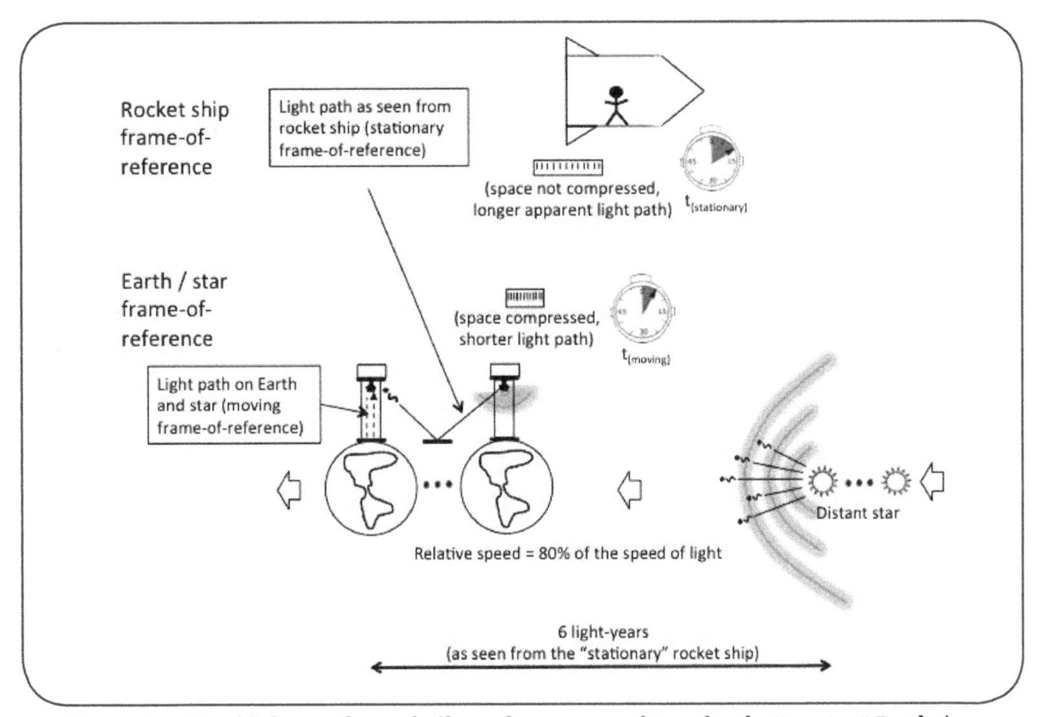

Figure 6-2. How high-speed travel affects the passage of time for the "moving" Earth / star. *In this illustration the Earth and star are moving at 80% of the speed of light relative to the rocket ship so the elapsed time on the Earth and star is less as measured from the rocket ship reference frame.*

Considering the strange way that time and length change, based only on where the observer is located and the relative motion involved, one might ask: isn't there an actual length of things and a standard for time passage?

The answer is yes. This will be discussed in Chapter 9. We will also see later that time slowing and space compression occur when accelerations and gravity are present. However, in the case of acceleration and gravity, the time slowing and space compression are not bi-directional. Accelerations and gravity always slow time and shorten space. Further, when persons leave one frame-of-reference, to join another frame-of-reference, this will involve some acceleration. When this happens, any slowed time experienced while on first "moving" frame-of-reference, gets "locked-in" when on the second (common) “stationary” frame-of-reference. In this case, the people that were originally on the “moving” frame-of-reference will have aged less than people who were always on the “stationary” frame-of-reference. There's much more on this later in the book.

Chapter 7
Shortcomings of Special Relativity and the Equivalence Principle

This chapter addresses the following topics and questions:

7.1 Shortcomings of Special Relativity. *How does the centrifugal force in a spinning pail of water and other forms of acceleration limit the applicability of Special Relativity?*

7.2 The nature of acceleration and gravity. *What is the relationship between acceleration and gravity? Why do they feel the same? How do they interact and why?*

7.3 The Equivalence Principle. *How are inertial mass and gravitational mass equivalent? How can acceleration and gravity produce equivalent effects? How can they be combined?*

Shortly after developing his Special Theory of Relativity, Einstein realized that it had important limitations. While Special Relativity provided vital insights into the relationship between time, space, and the movement of objects through space, it was not able to account for some observable effects of accelerated motion, such as motion associated with rotating objects like a spinning pail of water. Many scientists were content to say that Special Relativity described the effects of non-accelerating ("uniform" or "inertial") motion on time and space, but motion involving acceleration was of a different nature and therefore governed by different laws. Einstein was unsatisfied with this view and felt that the same laws should be able to describe both uniform and accelerated motion. This led him to develop some revolutionary insights into how the Universe works and eventually to expand his Special Theory of Relativity to a more general and comprehensive theory, now called the General Theory of Relativity. This will be introduced in Chapter 8, and expanded upon in Chapters 9 and 10. In the process of addressing this shortcoming, Einstein made some important discoveries about the nature of acceleration and gravity.

7.1 Shortcomings of Special Relativity.

The centrifugal force in a spinning pail of water and other forms of acceleration limit the applicability of Special Relativity. This section explains why.

Both Galilean Relativity and Special Relativity are based on the premise that all motion is relative and can only be described in comparison to other objects, as described earlier. But this premise would also say, for example, that a spinning pail of water could in fact be spinning on its axis or, alternatively, the Universe could be spinning around the pail. Both are equally plausible explanations as shown in Figure 7-1.

But how then can we explain the centrifugal forces experienced in the spinning bucket if it were the Universe that is spinning? Similar arguments could also be made about the rotating Earth. Is the Earth spinning on its axis or is the Universe spinning around the Earth? In fact, those living in ancient times did believe that it was the Universe that was spinning. They believed that the Earth was at the center of the Universe and that indeed it was the Sun, stars, and entire Universe that were spinning around the Earth.

This Earth-centered view of the Universe was ultimately rejected when Astronomer Nicolaus Copernicus showed that the Earth and the other planets all orbited around the Sun. The pre-existing Earth-centered view could not adequately describe observed planetary movements. The Earth-centered view required elaborate explanations to account for the complicated motions of the Sun-orbiting planets as seen from the Sun-orbiting Earth. The Sun-centered view simplified and clarified our understanding of the observed planet movements through the sky.

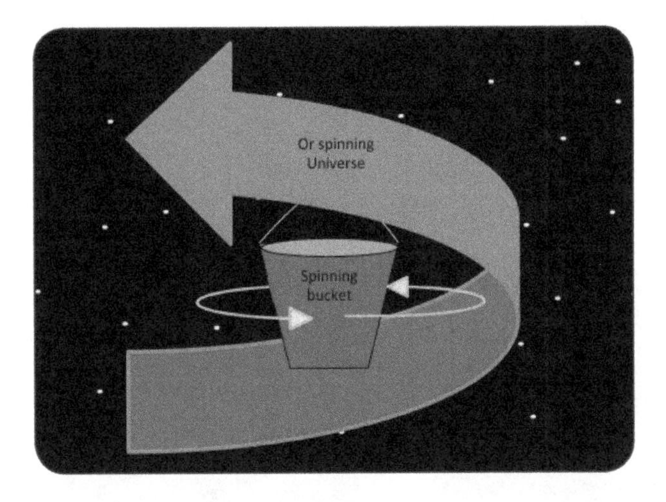

Figure 7-1. Is the bucket spinning, or is the Universe spinning around the bucket? *Could a "spinning" bucket just as easily be considered to be stationary with the Universe spinning around it in the opposite direction? If it is the Universe spinning around the bucket, then to explain the centrifugal forces experienced, Einstein concluded that motions involving accelerations are different from uniform motion and cannot be described by Special Relativity. He developed General Relativity to explain the centrifugal forces.*

The problem of explaining centrifugal forces led Einstein to conclude not all motion can be described simply by how objects move relative to each other. Laws that could also account for associated accelerations were needed. Specifically, he concluded that any motion that involves acceleration, like for a spinning pail of water, is different than uniform, non-accelerating motion. Any scientific theories describing motions that cause accelerations must be able to account for the accelerations created in addition to just describing the changing relative positions. And this means that motions involving acceleration cannot be described by Special Relativity. Special Relativity had to be expanded to also account for the accelerations.

For example, in the case of a spinning bucket, Special Relativity had to be expanded to account for the centrifugal force of the rotation; the force created by the continuous turning, or change of direction, of the walls of the bucket. This is the force that presses any liquid in the pail against the outside walls of the pail. Einstein expanded his Special Theory of Relativity to account for the centrifugal effects. His expanded theory, called the General Theory of Relativity, does this and is described in Chapter 8.

7.2 The nature of acceleration and gravity. In the process of accounting for the force experienced during accelerated motion, Einstein made some important discoveries about the nature of acceleration and its relationship to gravity.

Figure 7-2. Is the rocket ship accelerating through the Universe, or is the Universe accelerating past the rocket ship? *The acceleration forces felt on an accelerating rocket ship cannot be explained by the simple logic of Galilean or Special Relativity because these theories cannot account for the acceleration forces that are experienced inside the rocket ship.*

It was not just centrifugal accelerations that had to be addressed. The same questions arose for an accelerating object, like the rocket ship shown in Figure 7-2. Simply considering Galilean Relativity, one could say that either the rocket ship was accelerating through the Universe or that the Universe was accelerating past the rocket ship. However, if it was the Universe that was accelerating, then how do you account for the acceleration forces felt by passengers in the rocket ship? Again, Einstein discovered that any motion that experiences accelerations, even accelerations felt when driving a car around a curve, for example, needed new laws that can describe the motion as well as account for the accelerations.[1]

In both cases it's the acceleration that's at issue. Let's further explore the spinning pail of water, for example, to understand more completely why acceleration defines a unique class of motion. This exploration will also help us begin to understand why acceleration is a lot like gravity. Similarities between acceleration and gravity helped Einstein discover the true nature of acceleration and ultimately led him to develop the General Theory of Relativity.

First, consider the pale of water spinning on its axis in empty space outside the influence of gravity. The water in the spinning pail will be pressed against the sides of the pail and pushed out horizontally as shown in Figure 7-3. In fact, the occurrence of centrifugal accelerations on spinning objects in space has been recently demonstrated in experiments on the space shuttle. If an equally plausible explanation was that the Universe was spinning around the bucket, how could this happen? How could centrifugal forces (accelerations) be created, forcing the water out of the pail?

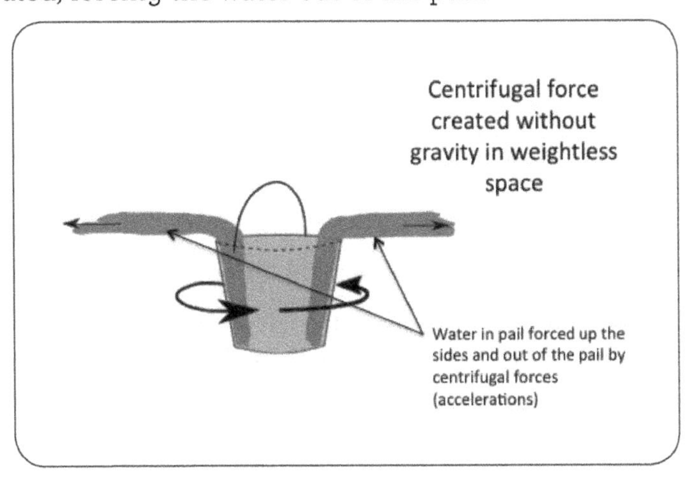

Figure 7-3. Spinning objects in weightless space experience centrifugal forces. *A spinning pail of water in space pushes the water to the outside of the pail walls and soon all the water will be forced out of the pail as shown.*

The analysis gets even more interesting when the spinning pail is on Earth. On Earth, the water would experience the Earth's gravitational force in addition to the centrifugal force generated by the spinning bucket. In

1 We will see in Chapter 8 that with General Relativity it is possible to select any frame-of-reference to be the "stationary" reference for comparing motion, even those that are feeling acceleration forces. However, it is more convenient and makes more intuitive sense to see the rocket ship as accelerating through the Universe.

this situation, the water would form a concave shape as the centrifugal force worked against Earth's gravity to raise the level of the water near the edge of the pail. Some water could spill out, but much of it will remain in the pail, held there by the Earth's gravity, as shown in Figure 7-4. The faster the pail spins, the more water will be forced out. This is in stark contrast with the situation in outer space depicted in Figure 7-3 in the absence of Earth's gravity. Earth's gravity appears to counteract the centrifugal force. The water in the spinning pail is forced out of the bucket, but is pulled towards the Earth.

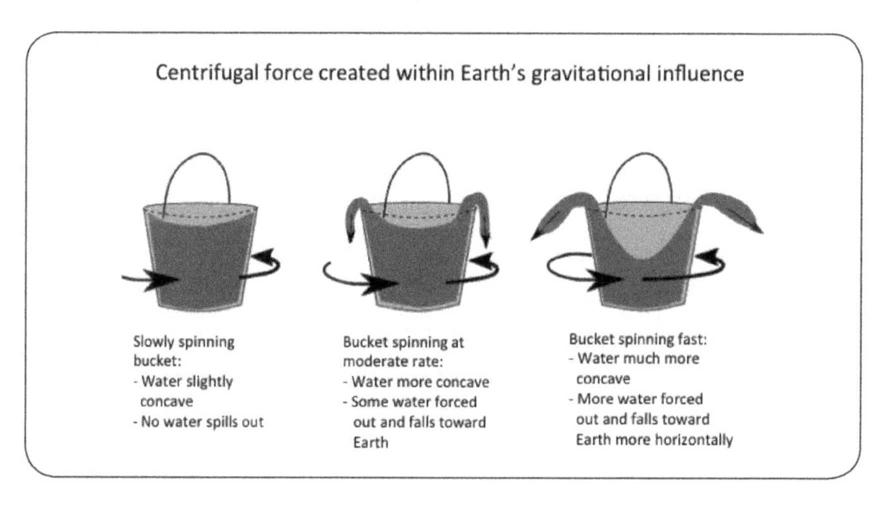

Figure 7-4. Interaction of centrifugal force and gravity. *Centrifugal forces of a spinning bucket combine with Earth's gravity to determine the amount of water forced out of the bucket.*

It was well known before Einstein that the effects of centrifugal acceleration were affected by the presence of gravity. It was always assumed that the effects of gravity mitigated, or worked against, any forces created by centrifugal acceleration. For the spinning pail example, the centrifugal force that would attempt to drive the water directly out of the pail in a horizontal direction (like in free space shown in Figure 7-3) would be partially overcome by the force of gravity causing the water to form a concave shape in the pail and to spill out in a direction angled toward the Earth. But Einstein came to a different conclusion as described in the next section.

7.3 The Equivalence Principle. Acceleration and gravity are the same force and both are related to the mass of the object involved. Inertial mass (associated with acceleration) and gravitational mass (associated with gravity) are "equivalent." This section defines these terms and describes how the effect of acceleration and gravity produce equivalent effects.

In considering the interaction between gravity and acceleration, including centrifugal accelerations, Einstein concluded that, rather than the force of gravity counteracting or working against the accelerations; the force of acceleration and the force of gravity are, in fact, the same force.[2] Let's consider Einstein's reasoning.

He came to this conclusion for several reasons. In addition to seeing the interacting effects of acceleration and gravity where gravity seems to moderate, for example, the amount of centrifugal force, Einstein also noticed that the sensation of acceleration is indistinguishable from the sensation of gravity. For example, traveling in an accelerating vehicle (like a rocket ship or elevator) feels exactly like, and is indistinguishable from, being in a gravitational field like on Earth's surface. This is shown in Figure 7-5. In the absence of knowledge about whether you are in an accelerating rocket ship or in a rocket ship sitting on Earth's surface, you cannot tell whether the force pulling you to the bottom of the rocket ship is due to the acceleration of the rocket ship or because the rocket ship is sitting in a gravitational field like that on Earth.[3]

2 Later in Chapter 8, we will see that acceleration and gravity are not really forces, but the result of the bending of space-time. However they feel like a force when experienced and are routinely referred to as a force or sometimes as a fictitious or pseudo force.

3 The amount of acceleration is sometimes quantified by the number of "g-s" present. The more "g-s" present, the more acceleration present. One "g" is equal to the amount of gravitational force felt on the surface of the Earth. If you were in a rocket ship accelerating at 2 "g-s," the force would be double the force of gravity on the surface of the Earth.

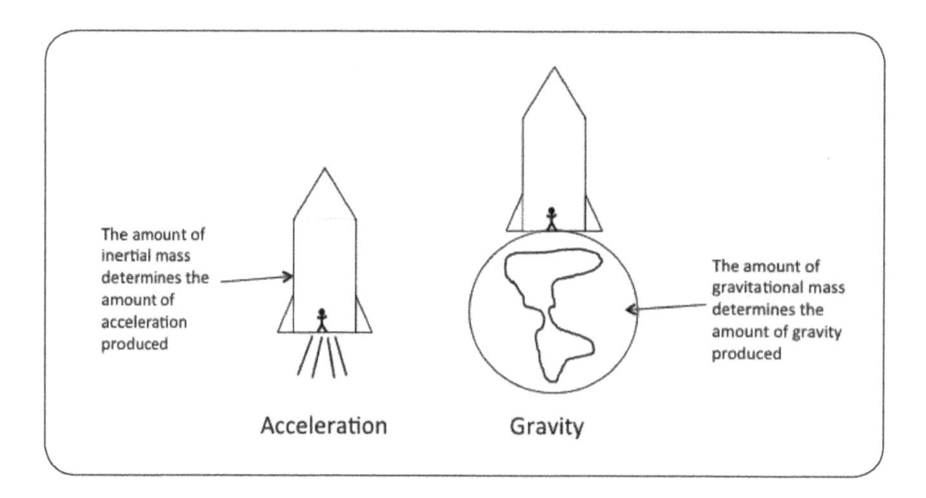

Figure 7-5. Gravity and acceleration are equivalent. *The force of gravity is indistinguishable from the force experienced when accelerating. Further, the amount of acceleration force felt in both situations is related to the amount of mass involved -- the mass of the object being accelerated in the acceleration case, and the amount of mass of the Earth in the gravity case.*

Not only are the sensations of acceleration and gravity indistinguishable, both are related to the amount of mass present.

- For the case of acceleration, a larger, more massive object experiences less acceleration for a given amount of force, while a smaller, less massive object experiences more. It takes more force to accelerate, or change the direction of movement, for a more massive object. A more massive object is said to have more inertia because it takes more force to cause it to accelerate, for example, to change its direction of movement. That is, the amount of mass an object has determines the amount of acceleration that can be achieved for a given amount of force. This is called the object's ***inertial mass***. Figure 7-6 shows an example of this relationship.

- For the case of gravity, the amount of gravitational force generated is determined by the amount of mass of the object producing the gravity. For example, the larger, more massive Sun generates 28 times more gravity than the Earth on their respective surfaces. The amount of mass associated with a gravity producing body is called an object's ***gravitational mass***. Because the Sun has more gravitational mass, it generates more gravitational force.

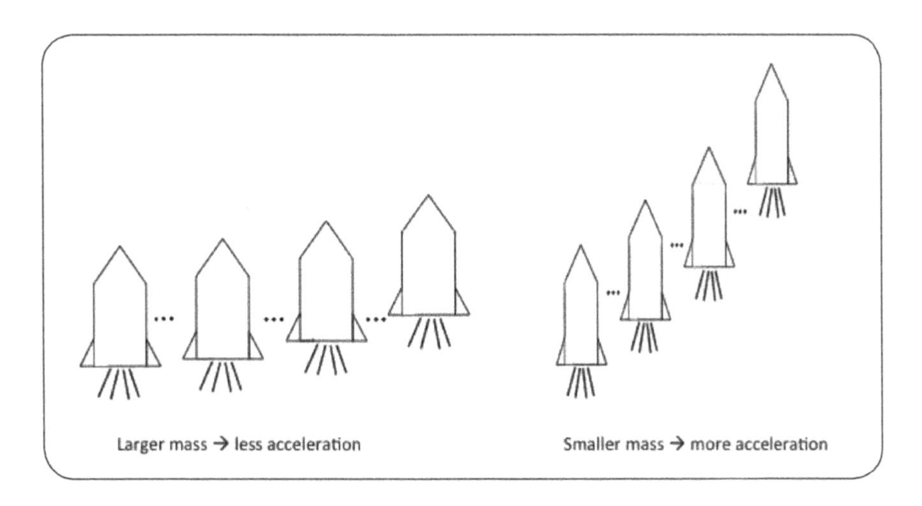

Figure 7-6. A more massive object is harder to accelerate. *The amount of acceleration that can be generated by a given jet engine is determined by the mass of the object being accelerated. The larger, more massive rocket ship on the left is accelerated less than the smaller, less massive rocket ship on the right, given the same jet engine (i.e., the same amount of force).*

In addition to the fact that both acceleration and gravity are related to an object's mass, Einstein showed that ***gravitational mass*** is actually equivalent to ***inertial mass***. That is, an object's mass determines the amount of gravitational force it creates and the amount of force needed to generate an equivalent amount of acceleration; and these forces are exactly the same for a given mass. This is called the ***Equivalence Principle***. The Equivalence Principle provides an elegant explanation for why all objects, regardless of their mass, fall toward Earth at the same rate (ignoring any effects of air resistance).

The Equivalence Principle explains why all objects fall to Earth at the same rate. This is because an object's higher inertial mass makes it harder to accelerate, while the object's higher gravitational mass creates more gravitational force. The effects are offsetting so that when two objects of different mass are dropped, they always fall at the same rate (ignoring any effects of air resistance). The reduced acceleration toward Earth achieved with the larger, more massive object is exactly offset by the larger object's greater gravity. Of course, Earth's gravity is the same for both objects, so it is the inertia and gravity associated with the falling objects themselves that offset each other. The increased gravity associated with the more massive object exactly offsets the higher force needed to accelerate it. All objects therefore fall at the same rate regardless of their mass.

To illustrate, consider the differential masses (gravitational and initial) associated with the balls in Figure 7-7. The larger gravitational mass of the larger ball creates more gravitational force, causing it to move toward the Earth more quickly. However, this larger ball also has more inertial mass, making it more difficult to accelerate. The two effects exactly offset each other. As Galileo famously is said to have demonstrated, the two balls fall toward Earth at the exact same rate.

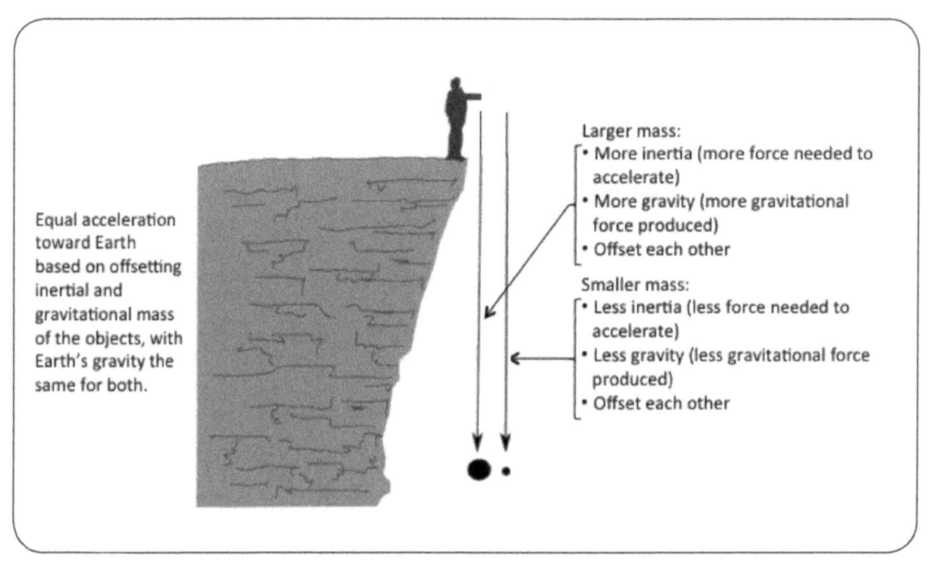

Figure 7-7. All objects fall at the same rate. *Many believe Galileo showed that objects of unequal mass fall at the same rate (ignoring any effects of air resistance). Einstein showed why: the more massive object creates more gravity, but also has more inertia.*

Since the force of gravity and force of acceleration are equivalent, then the two can be combined. We saw this with the spinning bucket. Figure 7-8 shows Einstein combining acceleration and gravity by riding his bicycle in a curved direction. By banking his bicycle during a turn, the force of acceleration is offset by the force of gravity, keeping the bike upright but banked. He is actually demonstrating that the two forces are equivalent by combining them. And by combining the force of acceleration and the force of gravity, Einstein has succeeded in changing the direction of "down." Is there anything he can't do!

Figure 7-8. Einstein combines gravity with acceleration. *Einstein showing us how the force of acceleration is the same as that of gravity and that when both are present they combine to change the direction of "down."*

Stop for a minute and feel the "pull" of Earth's gravity. Concentrate on how it feels; how it presses your body against the chair or ground. Later, when accelerating around a turn in your car or on a bike, feel how the sensation changes its direction, but not its nature. It still feels like the same force, but in a changed direction. The forces of gravity and acceleration are equivalent and can be thought of as the same force, just manifested through different mechanisms. The two forces are equivalent and are routinely combined.

It was Einstein who first realized that acceleration and gravity are the same force. This intuition ultimately led him to understand that both gravity and acceleration effect time and space in an equivalent manner adding to the effects of uniform motion described by Special Relativity. This formed the basis for the development of the General Theory of Relativity that expanded on the Special Theory of Relativity by also considering the effects of gravity and its equivalent acceleration. This is covered in the next chapters.

Chapter 8
General Relativity: Reality in Four Dimensions

This chapter addresses the following topics and questions:

8.1 Doubling down boldly. *How does General Relativity expand Special Relativity in a very bold way? How does it change our understanding of the Universe? What was its impact on science and physics?*

8.2 General Relativity requires a new kind of geometry. *What is it about our usual mathematics and geometry that doesn't work for General Relativity? Why was it necessary for Einstein to find a new kind of mathematics based on a different geometry in order to describe it?*

8.3 Space and time do not exist separately. *How are space (three dimensions) and time related and intertwined? How do they combine to form four-dimensional space-time? What does the combining of space and time into space-time say about the Universe?*

8.4 Space-time is distorted under conditions of gravitation. *What is really happening to space-time around sources of gravitation?*

Continue on page 76

Continue from previous page

8.5 Acceleration and gravity are equivalent but different. *How does acceleration differ from gravity? What is really happening to space-time during accelerations and gravity? Why can't we use our regular geometry to describe General Relativity?*

8.6 Motion and space-time distortions cause changes in light's appearance. *How do relative motion, acceleration and gravity affect the appearance of light? How do scientists use these effects to tell whether stars and galaxies are moving towards us or away from us? How can the light we receive from distant stars and galaxies tell us about the amount of gravity present?*

8.7 Extreme gravitational conditions create extreme time slowing and space compression. *How much does gravity distort space-time? What happens when extreme levels of gravitation cause extreme amounts of space-time bending? Do extreme levels of acceleration also bend space-time?*

Unsatisfied with Special Relativity's inability to describe the effects of gravity and acceleration, Einstein labored for over 10 years to expand on Special Relativity and broaden its applicability. The result was the General Theory of Relativity, which formed a new way of understanding how the Universe works. Special Relativity can provide approximate calculations for the effects of simple relative movements that do not involve accelerations or gravity. When accelerations and the effects of gravity are involved, the more complete formulations of General Relativity are needed.

The previous chapters described relativity using simple, familiar perspectives. This makes it easier to see how relativity works. However, these familiar perspectives can hide the deeper nature of relativity and the underlying structure of the Universe it defines. This chapter expands on these familiar perspectives and describes relativity in ways that try to capture its more complete, but less intuitive nature.

One familiar convention that helps us see how relativity works, but limits our full understanding, is that we think of space as a three-dimensional thing that we move through; and experience time as something completely separate from space with quite different characteristics and behaviors. We can control our movements through the three dimensions of space (within limits), while time seems to be a different kind of dimension that is completely outside of our control, always moving from past to present to future. Because we see time and space as different things, we usually describe relativity as something that slows time and compresses space, as done in previous chapters. While this is a correct description of what happens, it doesn't give a complete picture of how relativity and the Universe really work. A more integrated and accurate view is described by General Relativity and presented in this chapter.

Another human perspective that provides a more intuitive discussion, but limits our deeper understanding of relativity, is that we see gravity and acceleration as two separate and different things. We see gravity as a ubiquitous force that "pulls" everything toward the Earth, while accelerations occur when external forces are applied to objects. Accelerations do not seem to be associated with gravity. They seem quite different, even though they feel the same when experienced, and, as we saw earlier, when they interact with each other. The equivalence of gravity and acceleration described in Chapter 7 is not obvious to us. This chapter again reviews how gravity and acceleration are equivalent, and begins to relate this equivalence to how relativity works. It also discusses ways in which gravity and acceleration are different.

8.1 Doubling down boldly. Having broken the mold of physics with his Special Theory of Relativity, Einstein followed through with a far more stunning and complete theory of the Universe — the General Theory of Relativity. He boldly proposed General Relativity, a new and sweeping hypothesis that can describe the effects of all movements, including those involving acceleration, and those caused or influenced by gravity.

When conventional scientific understanding was not able to account for real-world observations, or for measurements taken of real-world phenomena, Einstein questioned the validity of those understandings. For example, in formulating Special Relativity, as described in Chapter 4, he said the reason that the speed of light doesn't change when we move toward or away from a light source is because it doesn't. The speed of light remains constant because time itself slows and space compresses when objects move relative to each other. And he developed the mathematics to back up these ideas.

While these were bold new ways of describing motion and light, they were based on ideas that other scientists were already beginning to explore. Scottish mathematical physicist James Clerk Maxwell's work on electromagnetism predicted that the speed of light (and the speed of all electromagnetic radiation) was fixed and unaffected by relative motion, even though he couldn't explain why. And Dutch physicist Hendrik Lorentz had already developed mathematical transformations that Einstein was able to use in building the mathematical foundations of his Special Theory of Relativity. Also, unknown to Einstein, French mathematician and theoretical physicist Henri Poincare had even predicted that there would soon be *"an entirely new mechanics"* in which no velocity can exceed that of light, and *"a principle of relativity according to which the laws of physical phenomena should be the same, whether for an observer fixed, or for an observer carried along in a uniform movement of translation."*[1] But Poincare and the others were unable

1 Henri Poincare's speech was reprinted in *Scientific Monthly*, April 1956.

to bring their ideas together in a cohesive, comprehensive, defensible and understandable theory. Einstein did with Special Relativity.

But while Einstein's Special Theory of Relativity built on emerging science to define a new physics of how time and space are affected by ***uniform*** motion, his General Theory of Relativity completely redefined how we understand the structure and nature of the Universe. It was an entirely new way of understanding time and space. While Einstein's Special Theory of Relativity brought together emerging thoughts and science in an understandable new way, the General Theory of Relativity was an entirely new science far beyond any current thinking at the time. Einstein was truly doubling down boldly when he proposed his General Theory of Relativity. It was Einstein's General Theory of Relativity that made him famous and changed physics forever.[2]

8.2 General Relativity requires a new kind of geometry.

Einstein had to find a new mathematical geometry in order to develop General Relativity.

One of the things that made the General Theory of Relativity so hard was that it required a new kind of geometry and a new complicated mathematics to represent it. The complexity of the math led Einstein to enlist help to work through it. Even with help from his colleagues, the original formulas that Einstein developed were extremely complex and difficult to understand. They were later simplified by other scientists, but still involve sophisticated mathematics and are well beyond the scope of this book. Those interested in Einstein's mathematical foundations for General Relativity, at a simplified level, are referred to in Einstein's very readable book, *Relativity: The Special and General Theory* (1924, reprinted 2007). It is not necessary to understand the mathematics of General Relativity to appreciate the beauty and insight about the Universe that it provides. This book proceeds down this less mathematical path.

One of the complications that made General Relativity mathematics so difficult is that the geometry we normally use for describing spatial

2 Einstein was eventually awarded the Nobel Prize in Physics for his contributions to science. However, this award was not for his discovery of relativity, which had not yet received widely accepted experimental verification data. He was given the 1921 Nobel Prize in Physics for his discovery of the photoelectric effect, which occurs when light strikes metal and can liberate electrons. This discovery helped build the foundation for quantum theory, describing how the Universe works at the subatomic level.

relationships (in three dimensions) does not work for describing the more complicated motions that occur in four dimensions. That is, the "Euclidian" mathematics that describes the three-dimensional geometry we are familiar with and studied in high school does not work for describing the dynamics of accelerating motions within four-dimensional space-time. Instead, Einstein had to find a different, more sophisticated mathematics that would work in a dynamic four-dimensional realm. What he found was a theoretical mathematics designed to describe complex surfaces that can bend and twist in many dimensions. This was mathematics developed by German mathematician Bernhard Riemann. Riemann's mathematics was very theoretical and had little real-world application before being adopted by Einstein as the mathematical foundation for General Relativity.

These Riemannian mathematics, as applied to General Relativity, work by defining "tensors" for each point in four-dimensional space-time. These "tensors" are bundled to describe the curvature for small regions of space-time. The mathematics was difficult to comprehend and work with, but perfect for describing the space-time of Einstein's General Relativity. The Riemannian geometry defines small regions of four-dimensional space-time in a way that specifies the curvature across each of the four dimensions.

Aside from the obvious complexities of describing motion through four-dimensional space-time using a point-by-point perspective, Riemann-based geometry also defines geometry that differs in important ways from the geometry we are familiar with. For example, Riemann's differential geometry ***does not require:*** (1) that all angles of a triangle add to 180 degrees; (2) that parallel lines never meet; and (3) that the circumference of a circle must be described as exactly two times the product of π times the circle radius *(circumference = $2\pi r$)*. These differences give Riemannian geometry the ability to describe General Relativity, but make it difficult to visualize within the geometric structures we are used to. And, as you might imagine, it is very difficult to visualize the workings of General Relativity by just looking at the math. As Einstein put it, the matter of relating General Relativity to the physical world *"lays no small claims on the patience and on the power of abstraction on the reader."* In other words, the space-time described by the Riemannian geometry of General Relativity is of a totally different nature than the geometry we are accustomed to, and therefore very difficult to visualize using our normal human perspectives. Riemannian geometry does not easily translate into a visual model or graphical representation we can readily relate to.

While it is very difficult to visualize the complex four-dimensional geometry defined by General Relativity, the following analogy helps illustrate the nature of the changes that happen when additional dimensions are added. This analogy describes the geometric effects of going from a two-dimensional world to one with three dimensions as a way of appreciating going from our three-dimensional space to the four-dimensional space-time. The analogy imagines a two-dimensional space defined by the surface of the Earth, ignoring Earth's curvature defined in the third dimension. This is shown in Figure 8-1.

In this two-dimensional space, we are only able to understand and move along the Earth's surface in the north-south and east-west directions. Without appreciating that the Earth's surface actually curves around a third dimension, we see space contained in just the two dimensions. We see the north-south lines (longitude) and east-west lines (latitude) as straight parallel lines continuing forever. And we see the intersections of the longitude and latitude lines as forming 90-degree angles. However, when we discover that there is a third dimension, and that the Earth's surface actually forms a giant three-dimensional sphere, we discover that the assumptions we made in the two-dimensional world were not true. Our north-south lines (longitude) and east-west lines (latitude) actually curve around and retrace their path in giant circles defined in the three-dimensions of the Earth's sphere. We also find that the angles between the longitude and latitude lines aren't actually 90 degrees, as we thought. This is analogous to what happens when we go from the three dimensions we are familiar with (north-south, east-west, and up-down), to the four dimensions of space-time (the three dimensions of space and the fourth dimension of time).

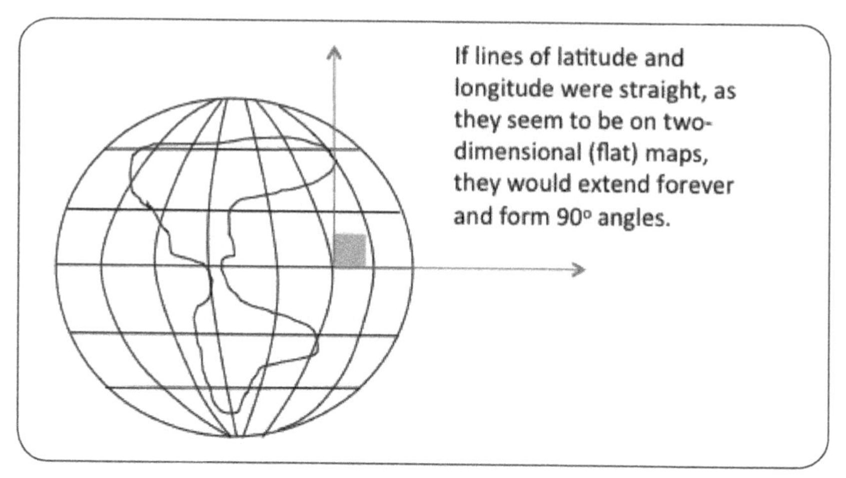

Figure 8-1. Adding dimensions affects geometry in unexpected ways. *Going from two to three dimensions creates geometric anomalies that cannot be easily predicted or understood when only considering two dimensions. This is similar to going from our accustomed three-dimensional space to four-dimensional space-time.*

Despite the mathematical complexities that form the foundation of our scientific understanding of General Relativity, it is possible to appreciate the basic principles of General Relativity without understanding or even seeing the math. It is possible to learn what General Relativity says about how our Universe works at a conceptual level that is completely accessible to non-scientists. Throughout this book, simple diagrams and analogies, like that in Figure 8-1, help us visualize and appreciate the principles of General Relativity without having to understand the math. These diagrams are not true representations of the four-dimensional reality described by the Riemannian geometry and mathematics of General Relativity, but they help us visualize and appreciate how they work.

8.3 Space and time do not exist separately. One of the clearest and most fundamental outcomes of Einstein's theories is the understanding that space and time are not separate things. They are part and parcel of one integrated thing, called space-time.

First and foremost of what Einstein's theories tell us is that time and space are not separate things; they are completely connected, intertwined and combined to form a single integrated entity that we experience, exist in and move through. They do not exist separately, but only in combination. When we move through space, we are actually moving through space-time; our movement has both a time and space component and the two cannot be separated.

When you think about it, this does not really go against common sense. It takes both a location and a time to say where we are or where we're going. For example, the statement that we would like to meet at the hospital on the corner of Main and Humboldt is incomplete without also specifying a time. And movement involves changes in both space and time — as we move from one spatial location to another, the movement takes a certain amount of time and we arrive at the new location at a new (later) time. We have changed our location in both space and time. That is, we have changed our location in space-time. It does not make sense to define a location without also specifying the time.

This is illustrated in Figure 8-2 that shows a city grid highlighting a trip from one city location to another along with the time expended during the trip. The figure shows both the space and time components of the trip highlighting how they are related. In the figure, time is expended during a trip through a city grid. At the point where the traveler is stopped at a red light, movement on the spatial dimensions is temporarily halted, resulting in changes occurring only along the time dimension during that period.

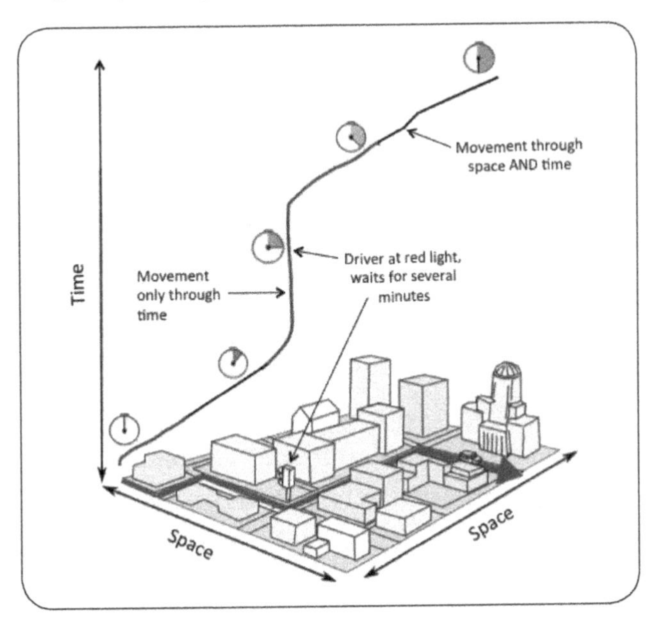

Figure 8-2. Seeing space and time as an integrated entity. *Locations in the spatial dimensions also require a time specification, and movements involve changes in both space and time. The close, integral relationship between space and time led Einstein to define them as integral parts of one thing that he called space-time (adapted from Gardiner, 1995).*

Einstein's theories have made the linkage and integration of space with time a defining aspect of how the Universe works. Not only are our movements within space tied to an associated time; when we move through space we are really moving through space-time. It's an integrated "Alice-in-Wonderland" reality in which our movement involves changes in both space and time together; that is, we change our position in space-time. And this affects how we experience time (when considered separately from space) and how we experience space (when considered separately from time). When considered separately, for example, time on fast, uniformly moving platforms seems slower and space appears to be compressed (as described earlier). These apparent changes reflect only changes in our perception of space and time when considered apart from integrated space-time. If we could experience space-time directly, in its integrated form, we would see these effects as simply a change of viewing perspective. That is, for example, seeing time from a different space-time angle, which is discussed further later in this book.

So how can we measure our movement through space-time in a way that respects the integration of time with space even if it is difficult to experience the integrated space-time directly? We know how to measure our movement through space; we measure the distances using familiar measures, like miles and feet and inches. And we know how to measure our movements through time; we count off seconds, minutes, hours, days, years, and so forth. But what about space-time? How can we measure our movements through space and time together? What measurements can we use?

The key to answering this question is based on the relationship between space and time. This relationship is 186,282 miles (space) per second (time).[3] Of course, this is just the speed of light. As noted repeatedly in the preceding chapters, the speed of light is the same regardless of any relative movements between objects (i.e., between frames-of-reference) involved. It stays the same because it defines the very relationship between space and time; it is how space and time are combined to form space-time. And for this reason, it provides the key for defining measures that we can use to gauge our movements through integrated space-time, even though our perceptions will focus on time and space separately.

3 As noted earlier, the selection of miles and seconds for units is arbitrary. We could just as well use meters and years or any other units for time and distance. For example, the speed of light expressed in meters and seconds would be 299,792 meters per second.

So what we do is measure our movements through space and time separately (as is natural for us), but select measurements that reflect and capture the integration between the two. Specifically, when measuring distance, we use measurements that incorporate the speed of light and thus capture the time component of space-time, as well as distance. Similarly, when measuring time, we use measures that relate time to distance, again by incorporating the speed of light. For example, we use light-years as a measure of distance; that is, the ***distance*** light travels in one year. And for measuring time, for example, we use a light-mile; that is the ***time*** it takes light to go one mile. Like for the light-year, the light-mile incorporates the speed of light into the measurement.

Einstein's theories of relativity integrate space with time to form space-time, and allow us to understand the effects of our movements through space-time. The Special Theory of Relativity defines the relationship of uniform motion through what we now understand is space-time. The General Theory of Relativity expands on Special Relativity to also consider the effects of accelerated (non-uniform) movements and gravity. Einstein did this by using a new mathematical geometry that defines and quantifies the integration between space and time. Using Einstein's formulas, it is possible to calculate both the space and time values (when considered separately) that result from relative movements through space-time, as well as the related distortions of space-time created by gravity and acceleration. These are described below.

8.4 Space-time becomes distorted under conditions of gravitation. Gravity affects how things move through space by affecting the "shape" or changing the "contours" of space-time. This can be thought of as a bending or compressing of space-time.

Gravity works by creating "force fields" that change the shape of space-time. It is similar to the "force fields" created by magnets, except that gravitational force fields change the shape of space-time itself. It is the bending of space-time that affects how the objects move, and even how light moves, when traveling through space-time. Larger, massive objects create stronger gravitational force fields that change the contours of space-time more than do smaller objects; space-time becomes more curved, more bent, around larger objects. It is the altered contours of "curved" space-time that affect the movement of things and light when viewed, considering only three-dimensional space and ignoring the time part of space-time.

Figure 8-3 uses a three-dimensional analogy to illustrate how gravity curves space in four-dimensional space-time. The figure shows how a gravity-producing object like the Earth creates spatial distortions on a two-dimensional surface, even though the actual nature of the distortions is only obvious when understood in the full three dimensions. This is analogous to how three-dimensional space is curved by gravity without appreciating the integration of the fourth space-time dimension, time. Objects seem to be "pulled" by gravity when seen in just the three dimensions of space. But if we could appreciate the more complete shape of four-dimensional space-time we would be able to see that the effect of gravity is not a mysterious force that reaches out and pulls on things, but is simply the result of changes to the shape of four-dimensional space-time. Gravity is not a "force"; it is the geometry of space-time. When objects follow natural straight-line paths through four-dimensional space-time they appear to be taking a curved path through three-dimensional space. This is an illusion created because we experience space and time separately and cannot see the shape or geometry of four-dimensional space-time in its integrated form.

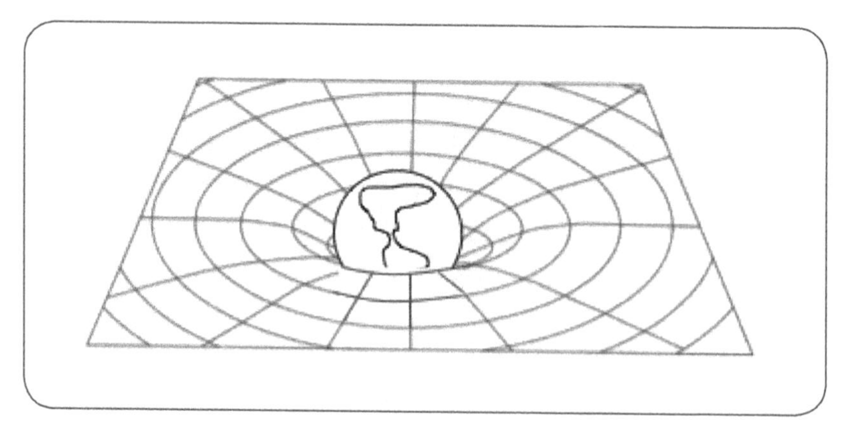

Figure 8-3. How gravity distorts space-time made obvious in a three-dimensional analogy. *One way to visualize how massive bodies like Earth distort four-dimensional space-time is to view a three-dimensional analogy like that shown here. In this analogy, the Earth distorts a two-dimensional space, creating natural contours only made obvious when seen in three dimensions. This is analogous to going from our three-dimensional space to four-dimensional space-time.*

So this analogy and visual representation illustrates the way gravity distorts the shape of four-dimensional space-time that we see made apparent in the three spatial dimensions. The analogy is used in most books on relativity. Through this analogy, we can begin to appreciate that gravity "attracts" objects by curving or changing the shape of four-dimensional space-time so that they appear to be attracted to the gravity source when seen in only three dimensions. While objects appear to "deflect" or be pulled toward the source of the curvature when seen in three dimensions, they are really just following the natural, "straight-line" space-time contours. It only looks like gravity attracts objects. The Earth's gravity feels like a force pulling us down because that's the direction we would go were we able to follow the natural contours of space-time; we would move toward the center of the Earth. It feels like a force when we stand on Earth's surface and defy the natural contours of space-time.

The difference between the apparent "attraction" of objects caused by an assumed gravitational force versus by a change in the contours of space-time is further illustrated in Figure 8-4. This figure compares a two-dimensional view (top portion) where an unexplained force is implied, but where the nature of how the force works is only obvious when the third dimension (bottom portion) is understood. In the top part of the figure when only two dimensions are understood or considered, an object like the Earth's Moon follows a path that seems to be curved and responding to an unseen and unexplained pulling force. However, when the third dimension is obvious, it can be seen that the Moon is just following the natural contours of the two-dimensional plane as curved in the third dimension. Using this analogy, we can understand that orbiting objects are simply following the normal, straight-line contours of curved four-dimensional space-time. These "curved" contours also affect the movement of light as shown in Figure 8-5.

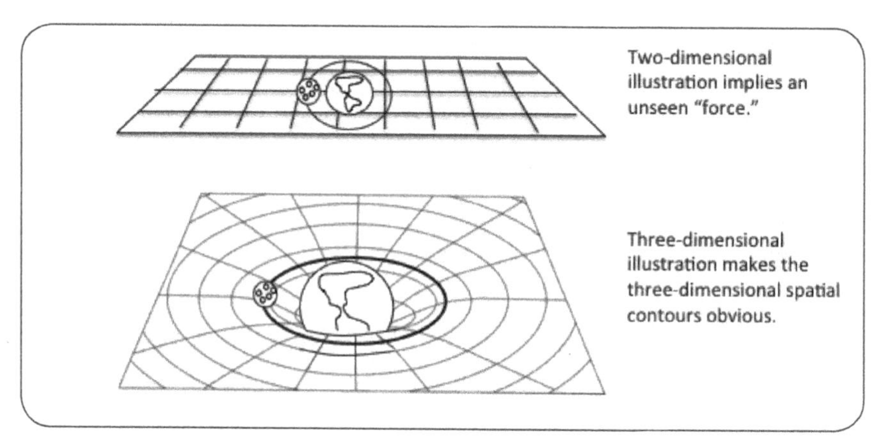

Figure 8-4. How the bending of space-time explains the otherwise mysterious "pull of gravity." *Without understanding the distortions visible in the third dimension, the orbital movement of the Moon, when understood in just two dimensions, would appear to be pulled by an unexplained force. This is analogous to the apparent "pull" of gravity when seen in just three dimensions without understanding the curvatures in four-dimensional space-time.*

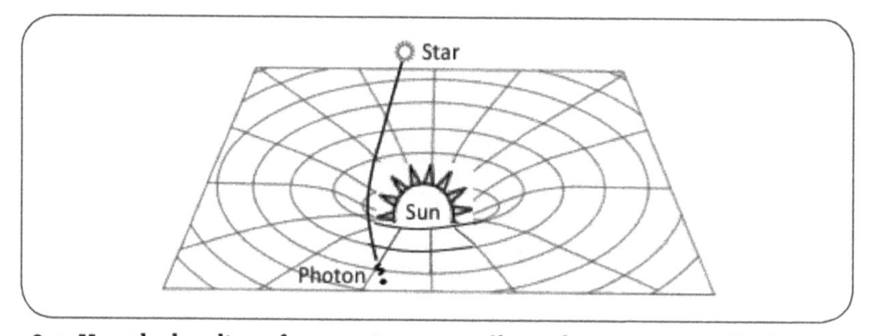

Figure 8-5. How the bending of space-time even effects the movement of light. *Space-time that is curved by the presence of a large gravity-producing body like the Sun even bends the path of light passing nearby. This was predicted by Einstein's Theory of General of Relativity and found to be true, thus validating the theory.*

The situation, simply stated, is that sources of gravity create distortions in space-time, which affect how objects move through space-time. It is these space-time distortions that make objects appear to take curved paths when viewed in three-dimensional space. This led physicist John Archibald Wheeler to explain *"space-time tells mass how to move; and mass tells space-time how to curve."*

8.5 Acceleration and gravity are equivalent, but different. Acceleration forces experienced on an accelerating rocket ship and those created by a gravity-producing body are equivalent but not identical. They are equivalent, but have some important differences, at least as we experience them.

For me acceleration is the most difficult aspect of relativity to get a handle on. Acceleration is not exactly the same as gravity, just a different manifestation of the same phenomenon as defined by the Equivalence Principle (Chapter 7). They feel the same, they are both related to mass in a similar fashion, and they can be combined like when water spills out of a spinning bucket in a gravitational field.

But there are a lot of questions that are difficult to answer. Does acceleration bend space-time like gravity? If so, how? How does the way acceleration bends space-time (if it does) relate to how gravitational fields bend space-time? Why do acceleration and gravity seem so different in how they work when we know they are equivalent? How do acceleration and gravity get combined? What happens to space-time when this happens? How does non-accelerating motion work in space-time that is curved by gravity? Complete and satisfying answers to most of these questions are hard to provide. Instead, a sense of what acceleration and gravity are, how they work, and how they are different will be discussed in this section.

Inertial and accelerating frames-of-reference. One key to understanding acceleration and gravity as we experience them is the distinction between "uniform motion," also called "inertial motion," and "accelerating motion."

Uniform (or inertial) motion is motion that does not experience any sensation of acceleration, such as those resulting from external forces like jet engines on rocket ships. External forces change the speed and / or direction of objects relative to space-time, and thereby produce the sensation of acceleration. With inertial motion, it is just the object's "inertia" that determines how the object moves. An object's "inertia," in the absence of any external forces or air resistance to slow it down, will keep on its present course and speed. It will just continue following the natural contours of space-time.[4] This is why it is called "inertial motion." For example, a car traveling 30 mph without any braking or acceleration will continue moving at that speed (ignoring the slowing that will occur from the friction from wheels and tires, and air resistance). A better example is an orbiting satellite. Orbiting satellites are truly in uniform motion following the natural contours

4 This is similar to Isaac Newton's First Law of Physics, which states that an object will remain at rest or in uniform motion in a straight line unless acted upon by an external force. Einstein refined this law with the understanding that motion is through space-time and that the natural contours of space-time can be curved by gravity. The notion that objects will travel in straight lines is no longer valid in Einstein's Universe since space-time itself becomes curved.

of space-time. They are not on Earth's surface fighting the contours of space-time that would pull them to the center of the Earth, and there is no friction or air resistance to slow them down. They just keep following the natural contours of space-time as curved by Earth's gravity.

Objects in inertial motion do not experience the sensation of acceleration, and are subject to the laws of Special Relativity. They are said to be in an inertial frame-of-reference, also called uniform motion. They can therefore be compared to other objects that are also in uniform motion using the laws of Special Relativity as described earlier. This is an important limitation of Special Relativity. Only frames-of-reference that are in uniform motion (not experiencing any acceleration "forces") can be compared using the laws of Special Relativity.

Objects that are ***not*** in an inertial frame-of-reference are objects that experience the sensation of acceleration due to changes in their speed or direction relative to the natural contours of space-time. The motion of these "accelerating" objects is referred to as "accelerating motion." They are said to be in an accelerating frame-of-reference. Because they involve accelerations, they require the more comprehensive laws of General Relativity to describe their motions and the accelerations felt. Special Relativity cannot be used for accelerating frames-of-reference.

Massive objects like the Earth bend space-time causing its natural contours to curve or distort. For the case of the Earth, these distortions bend space-time in a way that the curved space-time contours cause objects to move or deflect toward the center of the Earth, as described earlier. When objects near Earth follow the natural contours of space-time, they will appear to deflect toward the center of the Earth. Also, since they are in uniform motion, they will not feel any sensation of acceleration. With just the right amount of inertia and trajectory, objects in uniform motion can deflect around the Earth in a way that they continue orbiting around the Earth like the Moon and the many Earth-orbiting satellites. Orbiting satellites and the Moon are really in free-fall — no acceleration forces are felt — but their inertia keeps them aloft. By comparison, standing on Earth's surface we are not in free-fall, and are not following the natural contours of space-time as curved by Earth. We therefore feel the "forces" of gravity. This is illustrated in Figure 8-6.

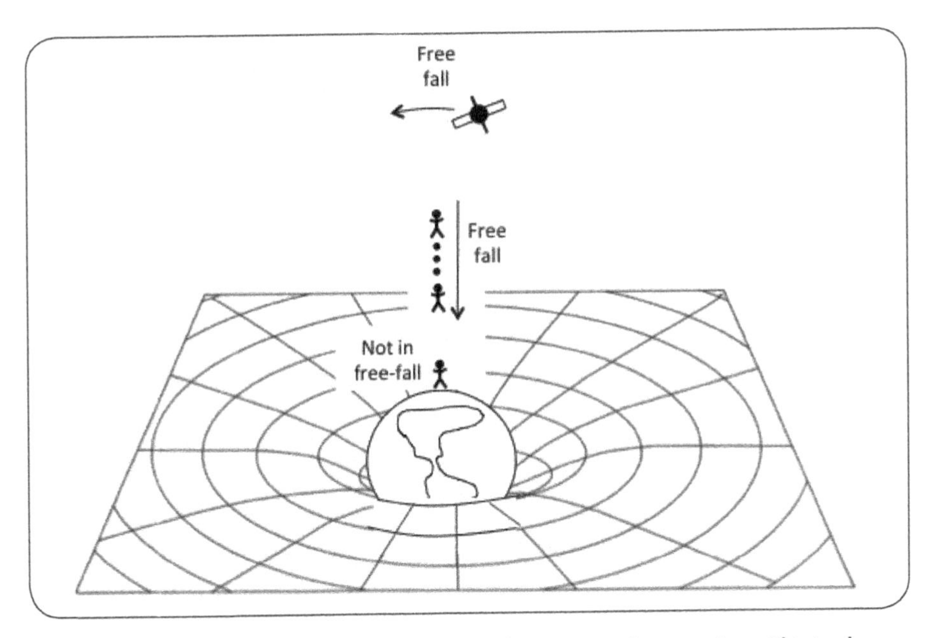

Figure 8-6. Free-fall means following the natural contours of space-time. *This is also called inertial or uniform motion. When we are on Earth's surface, we are no longer following the natural contours of space-time and therefore feel acceleration forces.*

Objects in free-fall toward the Earth are also in uniform motion, even though they approach Earth with accelerating relative speed. This is because they are following the natural contours of space-time as curved by Earth's gravity. Like for Earth-orbiting objects, objects in free-fall toward the Earth do not feel the sensation of acceleration. You can experience this when jumping off a diving board or parachuting before the parachute opens. You can also experience very briefly when in free-fall on a roller-coaster. All objects in uniform motion maintain their present speed and direction relative to the natural contours of space-time as curved by gravity, even though for the case of objects in free-fall and falling toward Earth, this means they are approaching the Earth with accelerating relative speed.

It is objects on Earth's surface that are no longer following the natural contours of space-time as curved by Earth's gravity, which would take them to the center of the Earth. They are therefore no longer in inertial motion. This is why we feel accelerations on Earth's surface.

To review, acceleration is felt when in an accelerating frame-of-reference, and not felt when in an inertial frame-of-reference. Accelerating frames-of-reference happen when energy or a structural barrier cause an object to deviate from the natural contours of space-time. This happens, for example, when a rocket ship fires its jet engines and when standing on the surface of the Earth. For the case of the accelerating rocket ship, observers feel acceleration forces when a rocket ship's jet engines force changes in speed and / or direction relative to the natural contours of space-time. For the case of standing on Earth's surface, it is the structural barrier of the Earth's surface that keeps objects from following the natural contours of space-time.

On the other hand, observers on a rocket ship that is not accelerating — either in space far away from sources of gravitation, or in free-fall toward a gravity-producing body — are in an inertial frame-of-reference. Observers in inertial (or uniform) motion do not feel any forces of acceleration. This is illustrated in Figure 8-7 for observers inside a rocket ship. This emphasizes that for a rocket ship accelerating at "one g" (same as Earth's gravity), it is hard (but not impossible, as we will see in a minute) to tell if the acceleration is from jet engines or by being on the Earth's surface.

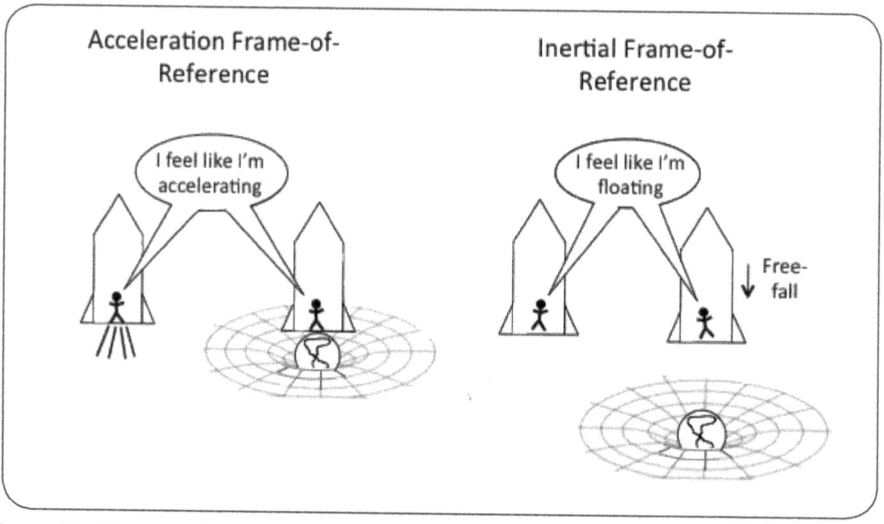

Figure 8-7. How gravity and acceleration feel. *The experience of being accelerated by a rocket ship and standing on Earth's surface produces the same sensation of acceleration. Both represent an accelerating frame-of-reference and are deviating from the natural contours of space-time. The experience of inertial motion involves an inertial frame-of-reference and is not associated with the sensation of acceleration.*

How accelerations from Earth's gravity and a rocket's engines differ. There are some important differences between accelerations felt from Earth's gravitational field, like when standing on Earth's surface, and accelerations felt from an accelerating rocket ship, even though these two situations feel the same.

One difference is the shape of the apparent gravity field creating the acceleration forces that are felt. The shape of the gravity field around large massive objects like Earth is circular, while the force created in the accelerating rocket ship, which can also be considered a gravity field (as we will see in a minute), is straight and pulls in just one direction opposite to the force of the jet engines. This results in small differences in the direction of the apparent forces felt. The force felt on the Earth seems to "pull" things angled toward the center of a big circle (the Earth) compared to the rocket ship where the forces are "straight back" in the opposite direction of the jet engine's force.

To illustrate, consider a person on an accelerating rocket ship dropping two balls held at arm's length. The balls will "fall" in a parallel fashion in accordance with a linear (straight line) force of the jet engines. However, considering the similar case of two balls falling toward Earth (or toward any massive gravity-producing body), the balls fall toward the center of the massive body and therefore get closer as they fall. This is called the gravitational tidal effect. On Earth, these tidal effects are very small and not noticeable. This is illustrated in Figure 8-8.

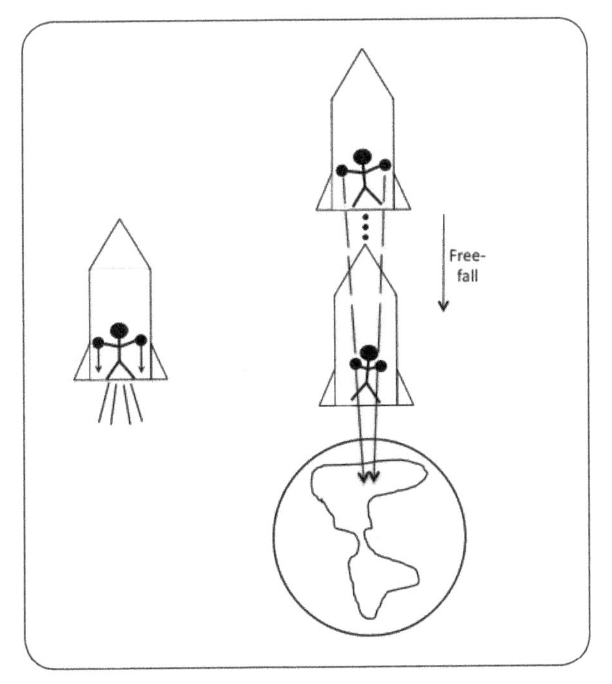

Figure 8-8. Accelerations created by accelerating objects and those due to gravitational fields feel different. *Accelerations created on accelerating objects like rocket ships are straight and in a direction opposite and proportional to the force of the rocket's jet engine. The accelerations felt on massive gravity-producing bodies like Earth angle toward the center of the Earth and are stronger nearer the Earth. This causes "falling" objects to behave differently from objects being accelerated on a rocket ship as depicted here.*

Another distinction between the two situations is that the field around the Earth is stronger nearer to the Earth, while the acceleration force from the rocket ship is constant and in one direction, opposite and proportional to the forces of the ship's rockets. The balls falling towards the Earth therefore experience a stronger gravitational field (space-time is more curved or compressed) as they get closer to the Earth and thus have a corresponding increase in the rate of speed change as they approach. That is, an object's relative speed increases at a faster and faster rate, the closer they get to the massive body where the "gravitational force" field (space-time distortion) is greater. For the case of the rocket ship, the force does not change based on location and is always proportional to the force created by the rockets. These effects are illustrated in Figure 8-8.

Note that for any object in free-fall and moving toward the Earth, the free-falling object and the surface of the Earth are ***both*** accelerating toward each other. However, only the object on Earth's surface is in an accelerating frame-of-reference and feels acceleration forces. The free-falling body is in an inertial frame-of-reference, as described above, and does not feel any acceleration forces. Nevertheless, they are both moving toward each other with increasing relative speed. This is because all motion is relative and defined by the changing relative positions with other objects.

In this example, Earth's surface, where acceleration forces are felt, represents a non-inertial (accelerating) frame-of-reference by deviating from the natural contours of space-time as curved by Earth. The free-falling object does not feel acceleration forces and is in an inertial frame-of-reference (associated with uniform motion). Accelerations that produce the sensation of acceleration are called "proper accelerations." Accelerations experienced on objects in free-fall toward Earth that do not produce the sensation of acceleration are called "coordinate accelerations."[5]

The larger context of relativity. The full beauty of General Relativity is that all frames-of-reference can be considered to be "stationary" and the effects of relativity considered from the perspective of each and every observer. This means that occupants on the accelerating rocket ship can consider themselves to be stationary with the acceleration forces felt assumed to be generated by

5 As described in Chapter 7, the Earth and objects falling toward Earth BOTH create gravitational fields and actually fall toward each other. However, since the Earth is so much larger than other things on and near Earth, the Earth's gravitational field overwhelms the other gravitational fields and all the motion is perceived to be toward the Earth.

a gravitational field from outside of the rocket ship. In fact, the rocket ship occupants may not know whether the rocket ship is sitting on Earth's surface, within Earth's gravitational field, or accelerating through space-time (with forces that equal one "g"). The laws of General Relativity work from either perspective.

This is one of the most profound results of Einstein's Special and General Theories of Relativity: *All motion is relative!* There is no preferred observational perspective (frame-of-reference), even though some perspectives make much more intuitive sense and are easier to imagine. For example, it is easier to think of the rocket ship as accelerating through space-time than to think of the entire Universe accelerating past. However, it is possible to apply General Relativity regardless of the perspective (frame-of-reference) chosen as "stationary."

It is the frame-of-reference chosen as "stationary" within General Relativity that determines what we call the acceleration forces felt. If the object or person feeling the acceleration forces is selected as "stationary," the forces are called gravitational. If the larger Universe is chosen as "stationary," the forces are called acceleration. The distinctions between how the acceleration forces are experienced (e.g., the presence of tidal forces) will hold as appropriate for the situation, regardless of the frame-of-reference chosen as "stationary." If the accelerating rocket ship is selected as stationary, the forces are called gravitational, and can be thought of as coming from the entire Universe.

Gravitational field shapes. Gravitational and acceleration force fields can come in a variety of shapes, such as the circular gravitational field around large, massive objects like Earth and straight-line acceleration fields created on board accelerating rocket ships. Spinning objects like a rotating disk or spinning bucket also create forces that are circular, similar to those around the Earth, but with the associated acceleration forces pushing outward. In both cases the force fields are circular with tidal forces that move objects either apart (like for spinning disks) or closer (like around the Earth).
As noted above, these forces can be considered to be gravitational if the frame-of-reference experiencing the acceleration sensations is selected as "stationary," or acceleration if the larger Universe is selected as "stationary."

Because of the differences in gravitational field shapes, it is possible for "stationary" observers experiencing the acceleration forces to figure out the shape of the field. For example, the presence or absence of tidal forces can allow the observer to figure out if the gravitational field is around a large, massive body like Earth, from a rocket ship being accelerated by jet engines, or from the inside of a rotating disk. The shape of some gravitational fields and the associated tidal forces are shown in Figure 8-9.

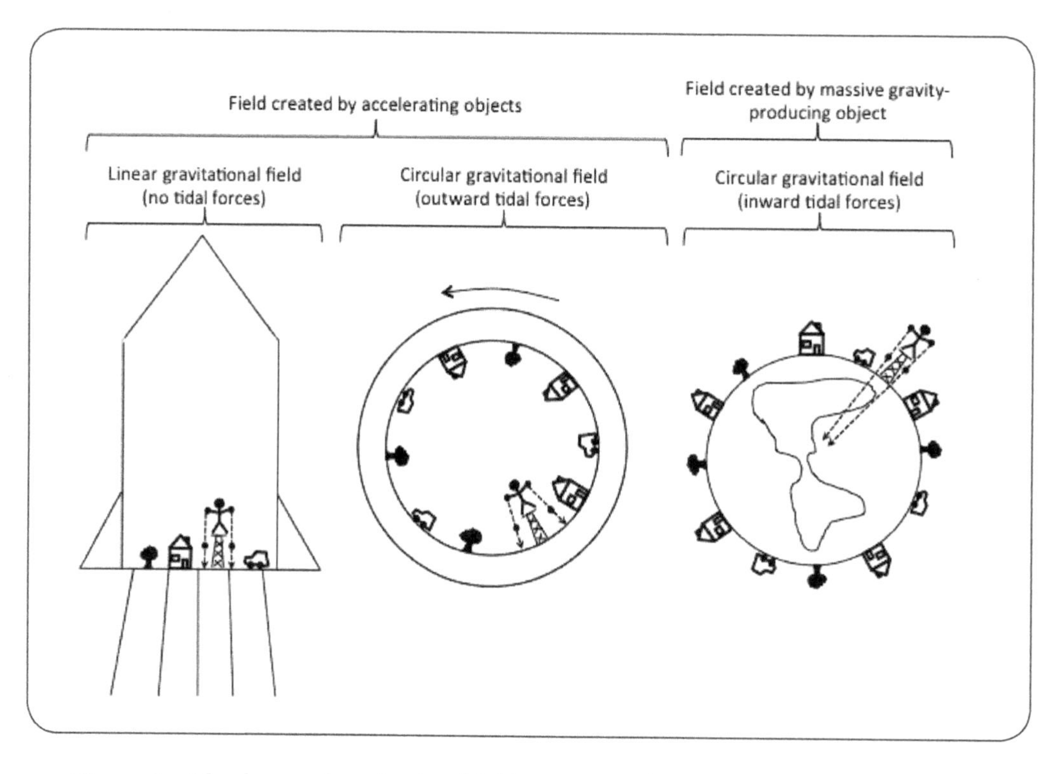

Figure 8-9. The shapes of gravitational fields can vary. *The proper acceleration forces felt on an accelerating rocket ship, a spinning disk, or on the surface of a large, massive object are equivalent and indistinguishable, except for the shape of the gravitational fields (the acceleration forces felt) that are experienced.*

In addition to the common types of situations where we experience accelerations, like those represented in Figure 8-9, there are some very strange gravitational effects that can happen around intense sources of gravitation, such as black holes. Black holes are stellar objects that are extremely massive, and have so much matter packed into a very small area, that the gravitational field is so intense that nothing can escape, not even light. When black holes are rapidly rotating they can create gravitational fields that deviate from circular ones. The application of Einstein's equations indicates that black holes, when initially formed, can rotate over 1,000 times per second with the resulting gravitational field shape significantly deviating from a circular shape, becoming more oval. The rotational Speeds of black holes are thought to slow over time, thus gradually changing the shapes of the gravitational fields to ones that are more circular in nature (like around the Earth and Sun, for example).

There are even more novel and strange gravitational effects that Einstein's equations predict may be possible when space-time is subjected to the intense gravitational effects of black holes. For example, two black holes in different locations in space-time, theoretically, can temporarily connect with each other. These theoretical connections between distant points in space-time are called wormholes and work by severely bending space-time to such an extent that the distant locations are temporarily connected. If wormholes are ever formed, it might be possible to pass through the wormhole, thus finding a shortcut across large expanses of space-time. In fact, wormholes are even stranger than this! Depending on the direction of travel through the wormhole, you may either move forward or backward in time. Wormholes are very theoretical and based on mathematical calculations, so no one really knows if they exist or exactly how they would work. They have not actually been discovered to exist and are thought to be very short-lived if they occur. See *Black Holes and Time Warps* by Kip Thorne for a nice discussion of these theoretical things and the strange behaviors they are thought to have.

Affect on the movement of light. Another difference between gravity and accelerating objects is their affect on the observed movement of light, the ultimate barometer for assessing the shape of space-time. First, let's look at gravitational fields created by large, massive bodies like the Earth. As we have seen, these gravitational fields curve space-time in a circular fashion around the gravity-producing body. This causes light to deflect and follow those curved space-time contours. Observers, regardless of their location and frames-of-reference, will be able to see these curved contours based on their effect on the movement of light. Observers will see light take curved paths as it traverses these curved areas of space-time, even though they may see it from different angles and from different frames-of-reference. They will be able to evaluate the amount of space-time curvature based on the path light takes going through it.

By contrast, acceleration forces caused by a rocket ship's jet engines create acceleration on the rocket ship itself as it continuously changes its frame-of-reference. This can be thought of as fighting against the natural contours of space-time. The rocket ship deviates from its prior inertial or accelerating state of motion, and this creates acceleration forces that are felt inside the rocket ship. The effect on space-time from accelerating objects like rocket ships is only experienced within the accelerating object, except in situations in which extremely large amounts of energy are involved. In situations involving extreme amounts of acceleration, space-time outside the accelerating object can actually become curved. This is because energy and mass are interchangeable, in accordance with the formula $E=mc^2$ discussed in Chapter 10. Since energy and mass are interchangeable, both can bend space-time. Therefore, the energy of the rocket ship's engine can theoretically bend space-time outside the rocketship. However, for the small energies of normal jet engines, the effect will be minimal. At extremely high energy levels, space-time outside the rocket ship can become curved or distorted.

For the case of acceleration by rockets, there is more happening than just accelerating motion. The rocket ship is moving through space-time. It is moving relative to other observing frames-of-reference, continually changing its location, relative speed and direction (frame-of-reference) relative to those observing frames-of-reference. The acceleration sensation (the proper acceleration) is evident on the rocket ship. Observers in other frames-of-reference see the effects of the rocket ship's acceleration as a dynamically changing frame-of-reference. The apparent slowing of time and compression of space simply due to Special Relativity will be seen as continuously changing as the accelerating rocket ship keeps changing its frame-of-reference. Those on the rocket ship feel a gravitational field and will see light "bending" within the rocket ship's frame-of-reference as the space-time contours are fought. Observers from other frames-of-reference just see a moving rocket ship continuously changing its relative course and speed. This is shown in Figure 8-10 from one selected frame-of-reference away from the rocket ship.

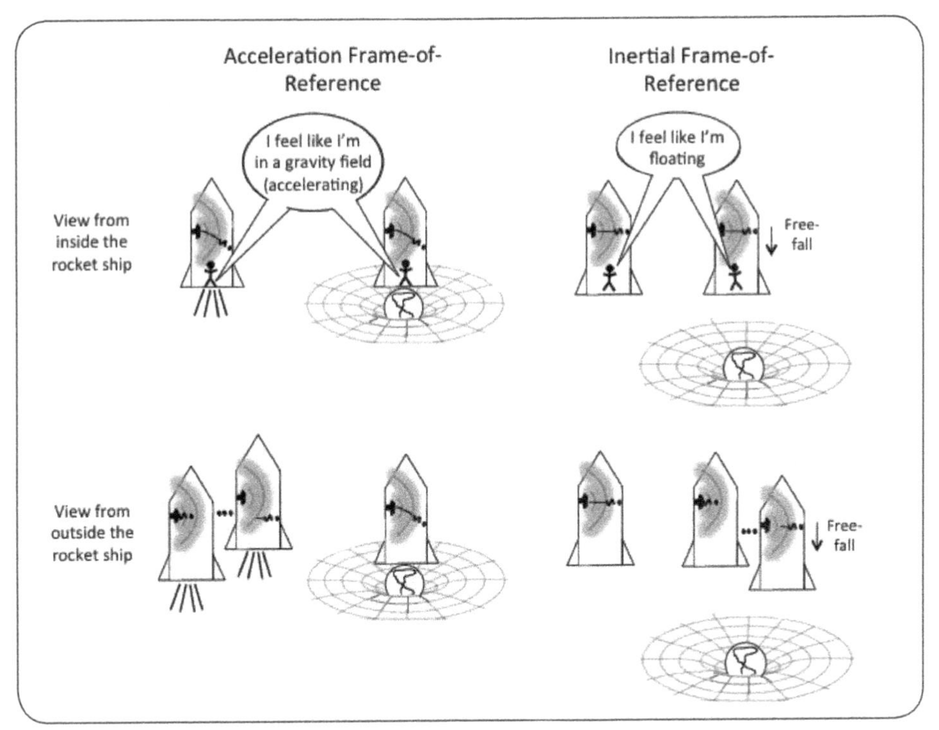

Figure 8-10. How gravity and acceleration affect the movement of light. *The movement of light across a rocket ship is affected by acceleration, which occurs when fighting the natural contours of space-time. This can occur in an accelerating rocket ship or when on the surface of a gravity-producing body, like Earth. The effect of acceleration on an accelerating rocket ship is experienced within the rocket ship while the bending of space-time by a gravity-producing body affects an area of space-time around the large body and is apparent to all observers.*

Why it's so hard to appreciate how relativity works. Throughout this book, simple two-dimensional diagrams have been provided to help readers see how both Special and General Relativity work. This necessarily leads to very crude representations of what is really happening. These diagrams help us visualize the effects of relativity, but don't come close to capturing the complex underlying reality, which requires a complicated "tensor calculus" based on Riemannian geometry developed by Bernhard Riemann. These simple diagrams are often the best we can do without digging into the math. They help us create a visual image of what is happening but can't fully capture the underlying complexities. Full and accurate representations can only be achieved with abstract mathematical representations.

While it is hard to visualize the four-dimensional workings of relativity, it is not hard to see why it is so difficult. Simply stated, it is because relativity works in four-dimensional space-time and uses a different kind of geometry with rules that differ from the "Euclidian" geometry we normally use for describing three-dimensional space. Einstein described a hypothetical situation for making the point more obvious. This is described below and illustrated in Figure 8-11.

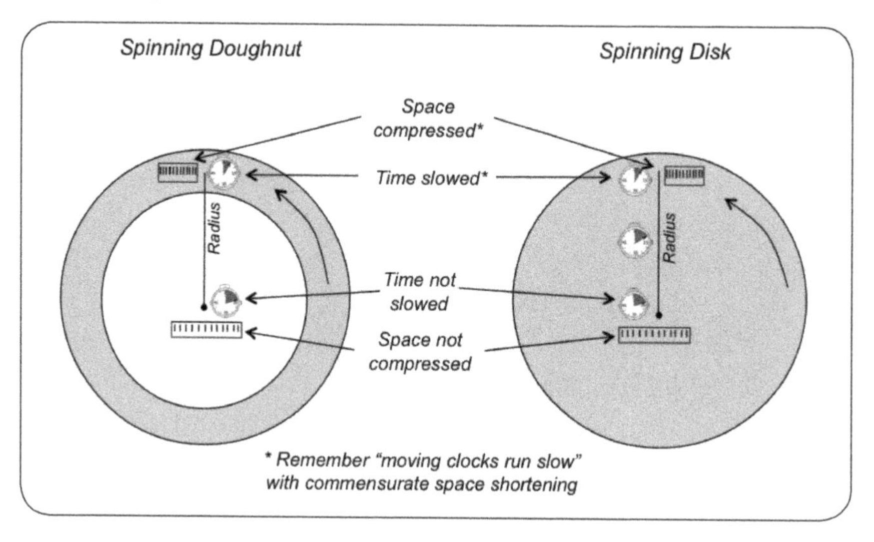

Figure 8-11. Why our normal Euclidian geometry doesn't work for relativity. *The outside of a spinning disk (solid or doughnut) is moving relative to the center. This means that time on the outer edge of the disk is slowed with an equivalent shortening of space in accordance with Special Relativity. But these effects do not apply at the center or along the radius of the disk, thus violating the normal relationship between the radius and circumference of a circle.*

Einstein asked us to imagine a spinning doughnut with clocks at its center and on its periphery. As viewed from the stationary center (a "stationary" frame-of-reference), the outside of the spinning doughnut is moving (right to left at the top of the doughnut) relative to the center. A clock on this moving doughnut will therefore be seen as experiencing slowed time and compressed space in accordance with the laws of Special Relativity (moving clocks run slow with commensurate shortened space). However, the "stationary" radius of the spinning doughnut does not experience shortening since its length is not moving across the view relative to the center. This means that the formula for the circumference of the spinning doughnut using the normal three-dimensional geometry we studied in high school, *circumference* = $2\pi r$, no longer holds because the circumference is shortened due to Special Relativity and the radius is not. Therefore, our normal Euclidian geometry no longer holds.

What's a little harder to appreciate is that, according to Einstein, this relationship also holds between clocks at the center and outside edge of a solid spinning disk. For the case of a solid disk, clocks on the disk periphery will still experience time slowing and space compression as seen from the center even though both the center and periphery are turning. This is because the clock at the center of the disk has no velocity compared to clocks at its edge, even though the center is turning angularly with the outer edge. Einstein also notes that clocks placed along the radius of the disk will experience time slowing and space shortening to a lesser extent than the clocks on the periphery. Since clocks along the radius will also experience more acceleration the further away from the center they are placed, this helps us also appreciate the relationship between the accelerations felt and the amount of time slowing. See Chapter 23 of Einstein's book, "Relativity: The Special and General Theory," for Einstein's explanation.

8.6 Motion and space-time distortions cause changes in light's appearance. While the speed of light is always speed "c," its frequency (i.e., color) changes in response to relative movements of the light source and to the curvature of space-time due to gravity.

The ***speed*** of the light does not change under conditions of relative movement, acceleration, or gravity. As stated repeatedly in this book, the speed of light (through a vacuum) is always speed "c." However, the ***frequency*** of the light does change. You will remember from Chapter 2 that when light waves vary in frequency this changes the color of the light. Higher frequency light looks bluer while lower frequency light looks redder. You will also remember that light is just a small part of the larger spectrum of electromagnetic radiation that also includes, for example, frequencies associated with radar, radio and x-rays. The effect of motion, gravity, and acceleration applies to frequencies along all portions of the electromagnetic spectrum. Frequency shifts can even go from visible light to, for example, the radar or radio portion of the electromagnetic spectrum. It is also possible for emissions outside the visible part of the electromagnetic spectrum to shift into the part of the spectrum that is visible as light.

So let's see how this works. The following sections describe light frequency changes caused by: (1) the simple movement of the light source; (2) the light emitted by an accelerating light source; and (3) the light emitted from within a gravitational field.

First, let's look at the simple relative movement of a light source. As noted in Chapter 2, light color is determined by the spacing of the light waves. Light waves become more spread out when a light source is moving away and more compressed when the light source is moving toward the observer. This is called the Doppler effect, named after the Austrian physicist Christian Doppler who first proposed it in 1852. The Doppler effect says that if the light source is moving toward the observer, the frequency of the light will increase, making the light appear bluer (or even move into the ultraviolet, x-ray, or gamma ray portion of the electromagnetic spectrum). This is similar to the passing of a car with a siren. The siren will have a higher pitch (higher frequency) when the car is approaching and will have a lower pitch (lower frequency) when moving away. This is because the sound waves become more compressed (higher frequency) when the sound source is moving closer and become less compressed (lower frequency) when the sound source is moving away. This changes the pitch of the sound. While sound waves and light waves are quite different in how they work (sound waves require a medium like air to move through, while light can move through "empty space"[6]), the same effect of the relative motion is experienced with respect to the frequency of the light waves. Both light and sound will have a higher frequency when the light or sound source is moving closer. The changing light frequency changes the light's color; the changing sound frequency changes its pitch. A similar effect can be seen for waves on the surface of a lake. The effect of relative motion for waves created on the surface of a lake is analogous to the waves of light created by a moving light source (a star). This is shown graphically for both situations in Figure 8-12.

6 Recent research is finding that outer space is anything but empty. Scientists are discovering invisible material throughout space that they call "dark matter" and "dark energy." The existence of dark matter is made apparent only by its gravitational effects on visible matter. Dark energy is made apparent by its accelerating effect on the expanding Universe. These are described in recent books, such as *The View from the Center of the Universe: Discovering Our Extraordinary Place in the Cosmos* by Joel R. Primack and Nancy Ellen Abrams.

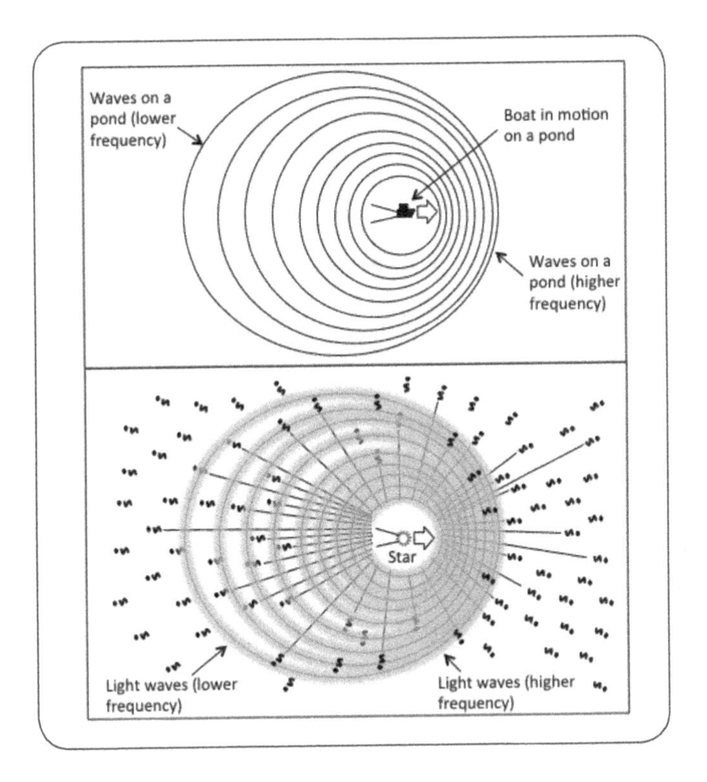

Figure 8-12. Like waves on a pond, the frequency of light changes with relative motion. *Waves on a pond created by a boat moving across a smooth water surface cause the waves in front of the boat to be compressed, thus having a higher frequency. This also happens for light waves when the light source is moving relative to an observer.*

The light (the photons) in Figure 8-12 is still moving at speed "c." It's just that the light waves will be seen by observers to have shifted in frequency due to the relative motion of the star. For example, the light will have a higher frequency when the star is moving closer to observers, making the light appear bluer in color. When the light source is moving further away, the effect is just the opposite: the frequency becomes lower and the light's color shifts toward the red end of the light spectrum, or to outside the visible range altogether. This is how scientists can tell if a star is moving towards us or away from us.

Now, let's look at an accelerating light source. The effects of acceleration of a light source on light's appearance are less easily visualized, but still affect light's appearance (frequency) in understandable ways. It's similar to simple movement of a light source, as described, except that the speed of movement for an accelerating light source is continually changing (e.g., increasing). The light source is moving faster and faster over time, so the effect of the relative movement keeps changing as the relative speed of the light source keeps changing. The frequency of the light, as seen by a "stationary" observer toward which the light source is accelerating, will continually increase, becoming less and less red (or more and more blue) in color. The left side of Figure 8-13 compares: (1) how light's frequency is higher for a light source that is moving at constant speed toward an observer with (2) how the light frequency is constantly increasing if the light source is accelerating toward the observer. The effect would be the opposite if the light source were moving or accelerating away from the observer.[7] For the case of a light source moving away, the frequency of the light will continually decrease (becoming more and more red) over time as the relative velocity of the light source continually increases in the away direction. Instead of moving faster and faster toward the "stationary" observer, it would move faster and faster away from the "stationary" observer.

Finally, let's look at the affect of gravity on light's appearance. Gravity has a similar, but not identical, effect as acceleration. For the case of gravity, light frequency gets lower as the light moves away from the gravity source, even though the light source is not moving with respect to the observer. This is because the geometry of space-time is less curved (think less compressed) further away from the gravity source. In this case, the frequency of the light depends on the amount of space-time "compression," which depends on the distance from the gravity source. Light emitted from an area of space-time whose shape is curved or compressed by gravity, will be seen at a lower frequency (than when sent) by "stationary" observers who are far away where space-time is less compressed. For example, if a blue-colored light were emitted from the surface of a massive gravitational source, like a large star, that light could appear green or even red to an observer far away from the star. This is because, while the speed of the light does not change, the spacing (frequency) of the light waves decreases as the "compression" of space-time decreases. This is called the "red shift" of light because lower frequency light appears more toward the red end of the visible light spectrum. In contrast to moving and accelerating light sources, where the effect on the frequency of light depends on the direction of the relative motion, the effect around gravity sources is always toward the lower frequency end of the spectrum (more red, less blue light) as you move away

7 The approaching light source example in Figure 8-13 lines up nicely with how light frequency changes around a gravity source to emphasize the similarities. However, for light sources that are moving away, the effect is the opposite. This is another difference between the affect of relative motion and gravity on the behavior of light.

from the gravity source, and this occurs in all directions around the gravity source. This is shown on the right side of Figure 8-13.

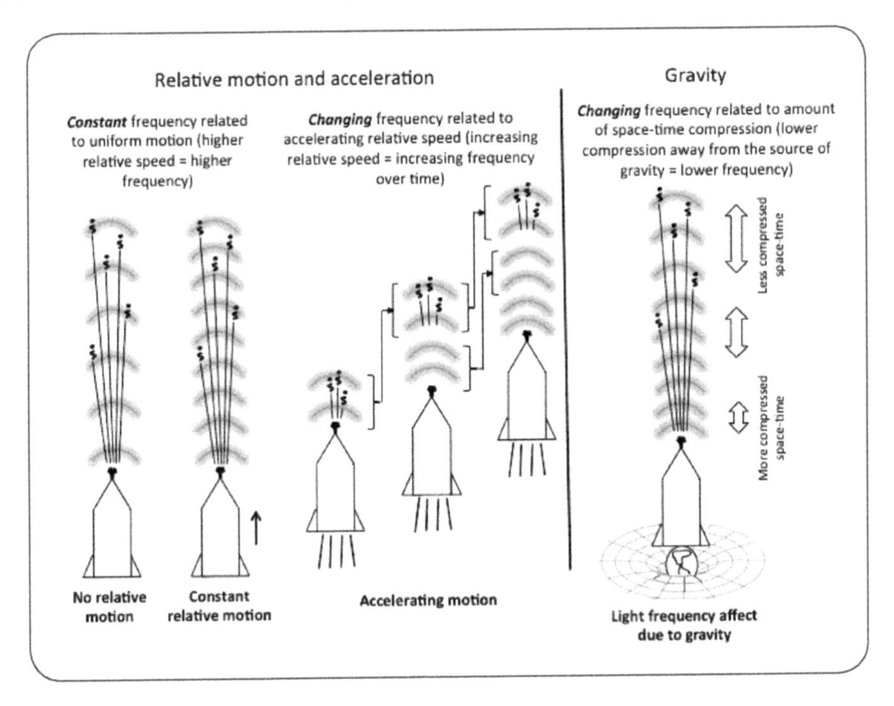

Figure 8-13. The affect of relative motion, acceleration and gravity on the color of light. *The relative motion of a light source changes the frequency of the light received by a "stationary" observer. In a similar, but not identical fashion, light emitted within a gravitational field will have a lower frequency when seen by an observer away from the gravitational source. In this illustration, the light source is at the nose of the rocket ship and the observer is at the top of the figure.*

To summarize, while the effects of acceleration and gravity on the appearance of light — that is, light's frequency not speed, which is always speed "c"— are similar, but they are not identical. This is shown in side-by-side comparison in Figure 8-13. Table 8-1 also provides a summary of these various motion and acceleration conditions on the appearance of light as seen by a "stationary" observer. Table 8-2 summarizes the appearance of light as affected by gravity based on the location of the observer. These tables emphasize that the speed of light remains the same regardless of light source movement or the presence of gravity.

Table 8-1.
Summary Table of Effects of Light Source Motion on Light Color

Condition	Light Source Motion	Affect on Light Appearance	Affect on Light Speed
No Relative Motion	Constant distance from observer	Constant light frequency	None, always speed "c"
Uniform Relative Motion	Moving toward observer	Higher constant light frequency (bluer color*)	None, always speed "c"
	Moving away from observer	Lower constant light frequency (redder color*)	None, always speed "c"
Acceleration Of Light Source	Acceleration toward observer	Continually increasing light frequency (bluer and bluer over time*)	None, always speed "c"
	Acceleration away from observer	Continually decreasing light frequency (redder and redder over time*)	None, always speed "c"

**Indication of redder or bluer is for general direction of changes in light frequency (i.e., toward that end of the visible light frequency spectrum that was described in Chapter 2). The actual color of the light observed depends on the color(s) of the light when emitted. Frequency shifts can even move into other ranges of the electromagnetic spectrum (e.g., radio waves on the low end to x-rays on the high end).*

Table 8-2. Summary Table of Effects of Gravity on Light Color

Condition	Location of Observer	Affect on Light Appearance	Affect on Light Speed
Gravity	Close to gravity source (e.g., on the sun's surface)	Same as when emitted	None, always speed "c"
	Away from the gravity source (e.g., far from the sun)	Lower frequency (redder*), reflecting less compressed space-time	None, always speed "c"

**Indication of redder is for general direction of changes in light frequency. For gravity, the direction of frequency shift is always toward lower frequency. The actual color of the light observed depends on the color(s) of the light when emitted. Frequency shifts can even move into other ranges of the electromagnetic spectrum (e.g., radio waves).*

As noted earlier, the actual way the frequency of light changes, as described by General Relativity, is much more complicated than shown on the two-dimensional picture in Figure 8-13. Again, this is because General Relativity is based on Riemannian geometry and associated complex mathematics. Nevertheless, the figure provides a way to visualize the effects.

What can the frequency of light tell us about the Universe? Because scientists understand how the frequency of light changes due to relative motion, accelerations and gravitational fields, the light we receive from distant locations in space provides important information for interpreting and understanding what's out there. For example, when we see changes in the frequency of light coming from space, we can make judgments about whether the light source is moving, the direction of the motion, and whether it is being affected by gravitational distortions of space-time.

Scientists are able to determine whether light frequency shifts have occurred by looking for known light frequency patterns associated with common chemicals like carbon, oxygen and water. These light frequency patterns are uniquely associated with those molecules and are very recognizable. It is possible to tell when these elements and compounds are present by finding the signature patterns. And scientists can detect relative motions, accelerations and gravity influences by seeing how the telltale frequency patterns are shifted in overall frequency. For example, if the pattern of light frequencies associated with oxygen molecules that usually appears between a given range of frequencies is shifted to a higher range of frequencies, we can assume that the light source containing the oxygen is moving towards us.

By recognizing known frequency patterns and analyzing how they are shifted from their normal frequencies, it is possible to determine the chemical composition and motions of the various stars, nebula and galaxies and how much gravity is present. Shifts in the frequency of light we receive from space allow us to map the movements of, and estimate the masses of, the stars and galaxies we see. This effect has been critical to our understanding of the Universe.

8.7 Extreme gravitational conditions create extreme time slowing and space compression. General Relativity predicted that extreme gravity sources would cause extreme space-time compression in which space is so compressed that even light cannot escape and time is slowed so much that it actually stops. Such an intense gravity source is called a "black hole."

General Relativity predicted that under extreme gravitational conditions, space-time becomes distorted to extreme levels. Such extreme gravitational sources are called "black holes." Black holes are places where so much matter is concentrated into an extremely small space, that gravity becomes so strong, and space-time so compressed, that even light cannot escape. For example, the black hole at the center of our galaxy, the Milky Way, contains a mass estimated at 4 million times that of the Sun. Under these extreme conditions, time actually stops, and light is trapped inside a small area surrounding the black hole, an area called an "event horizon." Anything that falls inside or past a black hole's event horizon is lost forever.[8] This is shown in Figure 8-14. Black holes are created when very large stars collapse after exhausting all of their nuclear fuel. Einstein's theories predicted the existence of black holes, but until recently there was no way to detect them. Today, we know that there are black holes at the center of many galaxies and elsewhere throughout the Universe. In fact, scientists have recently detected gravitational waves that were created when two black holes merged, which is also described later in the book.

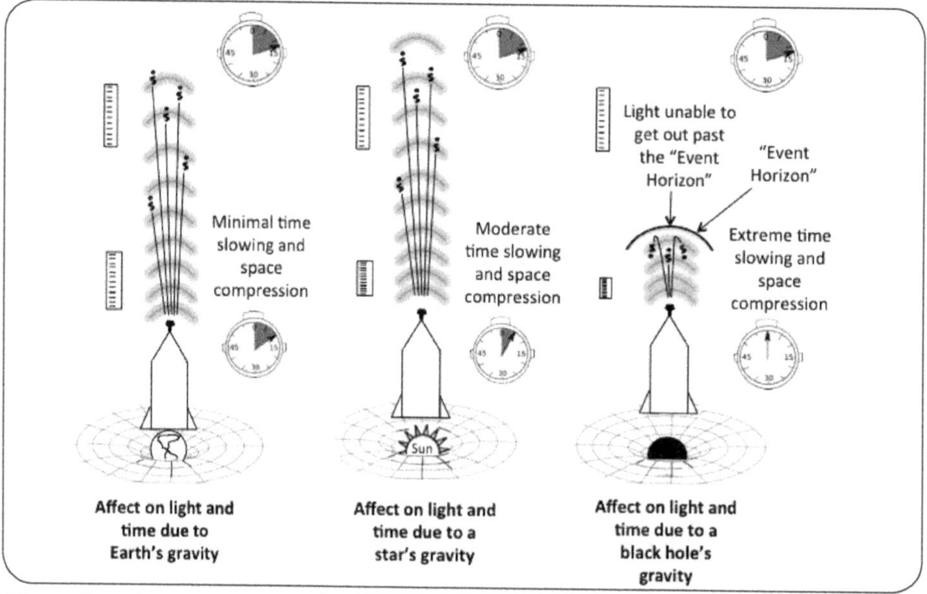

Figure 8-14. Large gravity-producing bodies distort space-time in a way that causes space to be compressed and time slowed. *Under extreme conditions, such as around "black holes," space-time is distorted to an extent where time stops and even light cannot escape as shown on right side of this illustration.*

8 Stephen Hawking recently presented scientific analysis that suggests that matter and energy inside black holes may be able to re-enter the normal Universe outside the "black hole" through a process called "black hole evaporation." This is a process in which the material in the "black hole" radiates to the outside Universe over large expanses of time. This radiation is called "Hawking Radiation." It is very theoretical and has yet to be observed. If this is true, "black holes" that do not attract in new material could eventually "evaporate" and disappear.

Chapter 9

What Relativity Says About the Universe: Part 1 — Time, Space and Aging

This chapter addresses the following topics and questions:

9.1 Time and aging. *Since relative motion, acceleration and gravity all affect the rate of time passage, will we age slower or faster depending on our relative motion? How did Einstein illustrate these effects even before experimental evidence supporting his theories could be developed?*

9.2 The universal speed limit. *Why can't we go faster than the speed of light?*

9.3 Simultaneous events and causality. *How can things occur at the same time if moving clocks run at different speeds? Is there really such a thing as simultaneous events? Can relativity change the order of events such that results can occur before the events that cause them? Why not?*

9.4 The relationship between event ordering, the slowing of time and the compression of space. *How does high-speed relative motion relate to the compression of space? Can we gain insight into why time slowing and space compression occur together by studying how high-speed motion changes the order of events in time?*

Einstein's theories have profound implications for how we understand space and time. This chapter presents important areas where our understanding of space and time have been shaped and changed by Einstein's theories of relativity.

9.1 Time and aging. Relative motion, acceleration and gravity affect the rate of time passage and can even affect how fast we age. This section describes the effects of relativity on aging using Einstein's twin paradox. It illustrates how relativity affects the rate of aging.

Both Special and General Relativity have the effect of slowing time. Does that mean we will age more slowly under differing conditions of relativity? Einstein said yes, and offered a "thought experiment," now known as the "twin paradox," to illustrate. This famous thought experiment is still instructive in helping us understand how relativity affects how we age and is presented here.

The twin paradox analytically describes how the effects of relativity would cause identical twins to age at different rates. According to the paradox, two identical twins are separated at birth and exposed to significantly different relative motion and acceleration conditions. Specifically, one of the twins remains on Earth, while the other is put in a rocket ship, accelerates to 80% of the speed of light, and travels to a nearby star that is 20 light-years away. Upon reaching the star, the twin then reverses course, decelerating then accelerating back to 80% of the speed of light, heading back to Earth. Upon her return, the traveling twin decelerates and rejoins her twin on Earth. This scenario is graphically shown in Figure 9-1. Note that decelerations are the same as accelerations with regard to the effects of relativity.

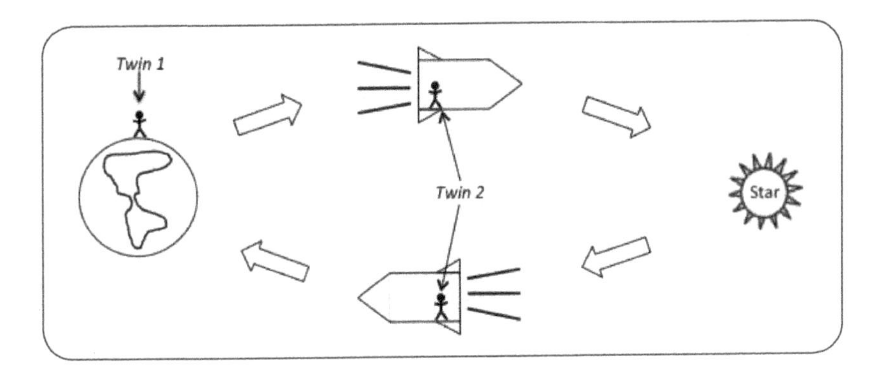

Figure 9-1. Einstein's twin paradox. *According to Einstein's "twin paradox," two twins are separated at birth. One twin is accelerated to near the speed of light and travels to a nearby star, turns around, and returns back to Earth. Einstein showed that the twin that traveled to the distant star would age less due to the accelerations and high-speed relative motion experienced.*

What Einstein showed with the twin paradox is that when you calculate the effect of high speed relative motion together with the accelerations experienced by the traveling twin, the twin that traveled to the distant star and then returned to Earth experienced slowed time from the perspective of the Earth-bound twin due to the effects of Special and General Relativity. This twin, therefore, aged less than the twin who stayed on Earth after they were both back in the same Earth frame-of-reference.

The conditions that caused the twin that traveled to the distant star and back to age less are summarized below:

1. Time moves more slowly on moving platforms due to Special Relativity. Since the rocket ship-bound twin traveled at 80% of the speed of light relative to Earth, this twin aged more slowly compared to the twin that stayed on Earth because of the relative motion, using Earth as the frame-of-reference selected as "stationary".
2. Acceleration has the effect of slowing time as shown earlier. The twin that traveled to the nearby star also experienced significant acceleration four times during the trip, causing time to go even more slowly during the accelerations.
3. At the end of the trip, both twins were in Earth's frame-of-reference. But the twin who experienced movement relative to the Earth — and accelerations during the trip — aged more slowly and was younger following the trip and reunification.

For this example, the twin that stayed on Earth would age 50 years during the traveling twin's trip. This is because at 80% of speed "c," it would take 50 years for the traveling twin to travel 40 light-years (40 light-years / 0.8 light-years per year = 50 years). The traveling twin, by comparison, given the relativistic time correction factor associated with traveling at 80% of speed "c" (this was calculated to be 60% in Chapter 6), would have aged 30 years (50 years x 60% = 30 years). Therefore, upon return to Earth the traveling twin would have aged 20 years less, just due to the relative motion alone (Special Relativity). The accelerations experienced would add to the age differences.[1]

But would similar results be obtained if the two twins simply flew past each other and never shared a common frame-of-reference? The answer is no! In the case of just speeding past each other, each twin would maintain her own frame-of-reference; they would never have a common frame-of-reference for absolute comparison. In this situation, each twin would think that the clocks for the other twin were moving slowly and each would think that they were experiencing normal time passage. And both would be correct. Unless one twin changes to the other's frame-of-reference, a change that would require some acceleration, they would never have a common frame-of-reference to experience an absolute time difference. They would simply fly past each other thinking the other's clocks were running slow due to the effects of Special Relativity. Which clock is running slow depends on which frame-of-reference you are measuring from. This situation of just flying past each other is shown in Figures 9-2 and 9-3, respectively for each twin.

1 This example ignores the slowing of time that the twin who stays on Earth experiences due to Earth's gravity. In fact, the time slowing due to Earth's weak gravity would be very small compared to accelerating to and from 80% of the speed of light four times during the trip. The twin paradox could just as well be carried out in space in the absence of any significant source of gravity.

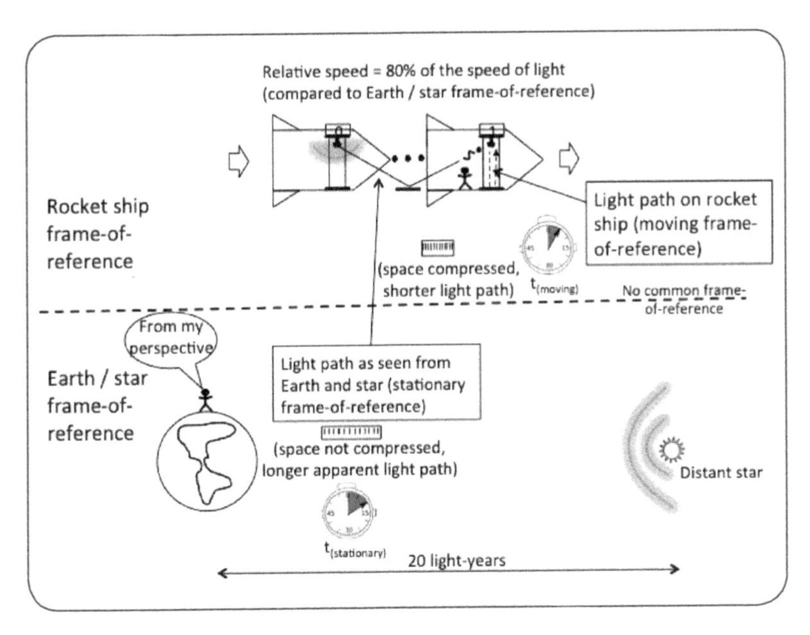

Figure 9-2. The twin paradox — Earth-bound perspective. *From the perspective of the twin on Earth, the clocks for the twin flying past would appear to be running slow.*

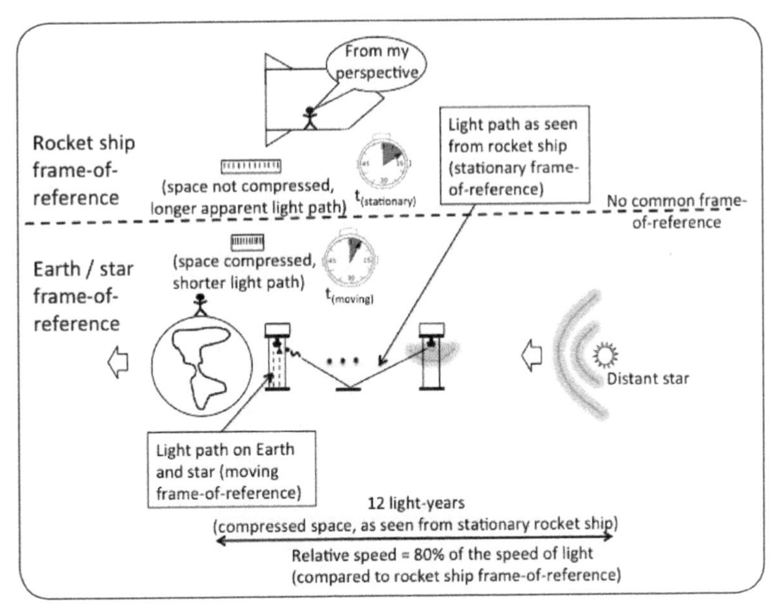

Figure 9-3. The twin paradox — spaceship perspective. *From the perspective of the twin on the spaceship, it would appear that the Earth and star were whizzing past and clocks on Earth were running slow.*

It's only when both twins make their judgments from a common (shared) frame-of-reference that they can agree and objectively compare the absolute differences of their time experiences. And in making the absolute comparisons, it will be the twin who experienced accelerations, in addition to the relative motion, who will be younger, when experienced in the final common frame-of-reference. It will be the shared perspective of the final common frame-of-reference that determines the effects of accelerations and relative motions that were experienced.

Another question worth asking is how is it possible for the traveling twin to reach a star that was 20 light-years away and return to Earth (40 light-years total) and only age 30 years? That is, how is it possible for the traveling twin to cover 40 light-years in only 30 years? As described in Chapter 6, the answer is that, in addition to experiencing slowed time, the traveling twin also experienced compressed space. In this case, the total trip length for the traveling twin was just 24 light-years due to space compression that results from Special Relativity (40 light-years x 60% = 24 light-years as calculated for a similar example on Page 57).

Yet another important question to ask is whether the slowing of time on moving platforms, or under conditions of acceleration or gravity, really affects real world processes, such as aging. That is, is it just moving clocks that run slow or are there real impacts on physical processes? Will people actually age at different rates depending on their relative motion? Will fires burn more slowly?

Today, we know that it is the deeper nature of time, and not just our measurement of time, that is affected by relativity. For example, the decay of radioactive material traveling at very high speeds has been shown to happen more slowly compared to identical radioactive material that stayed in the "stationary" frame-of-reference on Earth. This and other experiments, described later, have demonstrated real physical effects of time slowing. The two twins, when back on Earth together, will actually be different ages; the twin who stayed on Earth will really be 20 years older than her traveling twin, and you will be able to see the difference when they stand next to each other!

But the experimental results were not available during Einstein's time. It took experiments conducted after Einstein's death, to fully confirm both Special and General Relativity. Some of these experiments are described in Chapter 11. Instead, Einstein relied on "thought experiments" to illustrate his hypothesized results. Most famously, his twin paradox described here analytically showed how twins would age differently under differing conditions of relativity.

9.2 The universal speed limit. It is not possible to exceed the speed of light relative to any other frame-of-reference even when combining relative velocities. This section explains this principle.

One of the more difficult principles of relativity to appreciate is why it is not possible to go faster than speed "c." This seems to conflict with common sense and how we experience motion every day. For example, as described in Chapter 3, when we walk down the aisle of an airplane, our speed relative to the airplane would be about 2 mph, but would be about 502 mph relative to the Earth (assuming the plane were traveling at 500 mph). That is, the two velocities would add together as shown in Figure 9-4.

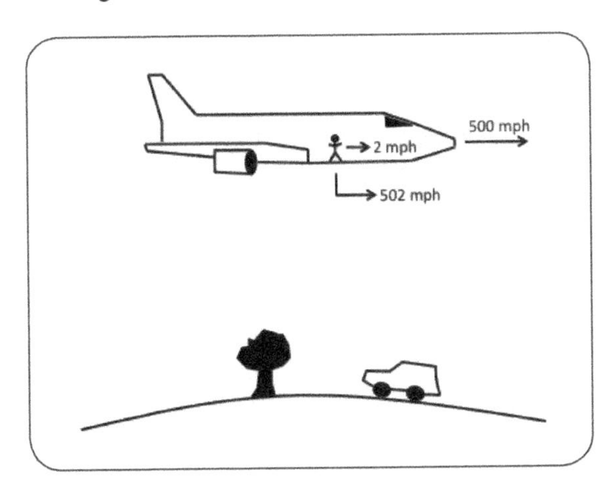

Figure 9-4. Adding velocities makes common sense. *When walking down the aisle of an airplane, your speed would be about 2 miles per hour relative to the airplane; but after adding the speed of the plane, it would be about 502 miles per hour relative to the Earth below.*

It would seem by extrapolation that if the plane were flying at 80% of the speed of light relative to the Earth and you were also walking down the aisle at 80% of speed "c" relative to the plane (imagining that this is possible), you would be moving 1.6 times speed "c" relative to the Earth. However, when we account for the effects of Special Relativity on time and space — that is the effects of slowed time and compressed space described earlier — the combined speed relative to the Earth is just 98% of speed "c." This is shown in Figure 9-5. The effects of relativity also apply to the lower speeds shown in Figure 9-4, but the effects on time and space are negligible at such low speeds; they are not significant so it is possible to simply add velocities and ignore the effects of Special Relativity. For those interested, the formulas and calculations for combining the relative velocities of these examples are presented in Appendix B.

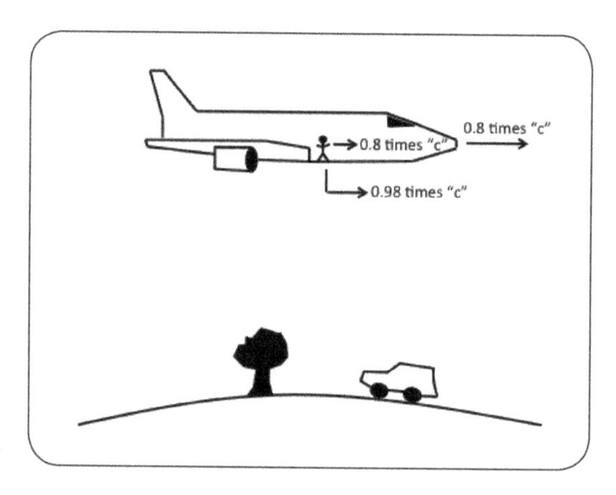

Figure 9-5. Adding velocities considering the effects of relativity (nested example). *If an airplane were able to fly at 80% of the speed of light and one could also walk down the aisle of the airplane at 80% of the speed of light, the combined speeds relative to the Earth would be 98% of the speed of light after accounting for the effects of relativity on time (time slowing) and space (space compression).*

Similarly, if two airplanes were flying toward each other at speeds near the speed of light, you would expect the two velocities to add together resulting in speeds above speed "c." This is shown in Figure 9-6. In this case, if both airplanes were approaching each other at 80% of the speed of light (relative to the Earth), like in the above example, you would expect their combined approach speed toward each other to be 1.6 times the speed of light. However, again applying the calculations Einstein developed for the Special Theory of Relativity that account for slowed time and compressed space, the combined velocities work out to be, again 98% of the speed of light (as shown in the example in Figure 9-6). This illustrates the universal speed limit! Those interested in the simple algebra calculating these speeds can find it in Appendix B.

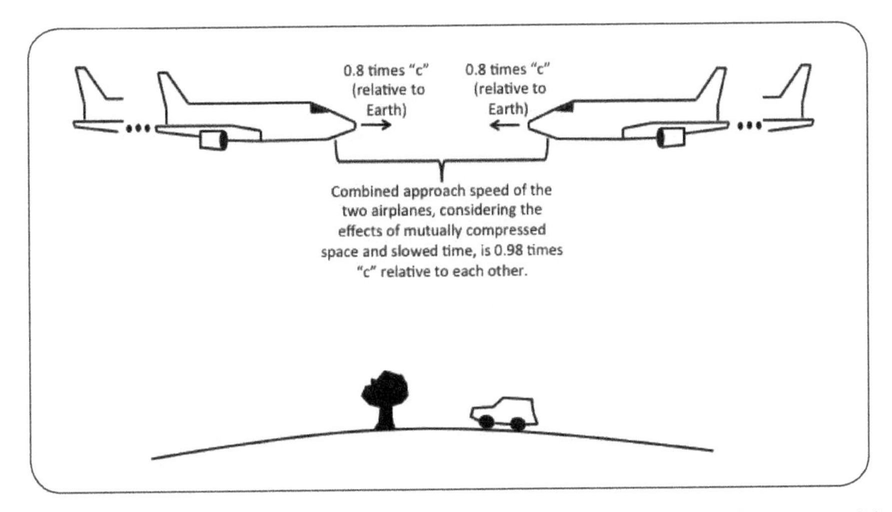

Figure 9-6. Adding velocities considering the effects of relativity (approaching example). *If two airplanes approached each other at 80% of the speed of light relative to the Earth, their combined approach speed would be 98% of the speed of light after accounting for the effects of relativity.*

But what if two airplanes were to approach each other at the speed of light? It turns out that this is not possible and therefore a meaningless question because any objects that have any mass at all cannot travel at speed "c" relative to each other or to anything else (i.e., compared to any other frame-of-reference). Any attempt to achieve speed "c" by anything with mass would require an infinite amount of energy and is not possible – this is one of the interpretations of the formula $E=mc^2$, which is described in Chapter 10. If two planes were to approach each other at 0.999 times speed "c" relative to the Earth, their combined speed relative to each other would be 0.999 times "c." Only massless particles, like photons[2], can go speed "c"; and photons, as we have seen, only and always go speed "c" (in a vacuum).[3] They don't go any other speed. This is shown in Figure 9-7. If two photons travel toward each other at speed "c", the combined approach speed will be speed "c." Appendix B contains these calculations.

2 Photons are considered to be massless. They are more like bundles of energy. However, they do have momentum and other characteristics like spin that we normally attribute to objects that do have mass. Physicists actually describe photons as having no "intrinsic" mass and therefore able to travel at speed "c;" but photons have "effective" mass by virtue of their energy and the relationship between mass and energy as defined in $E=mc^2$.

3 As noted earlier, light travels at speed "c" in a vacuum and somewhat slower when it goes through other mediums, like water.

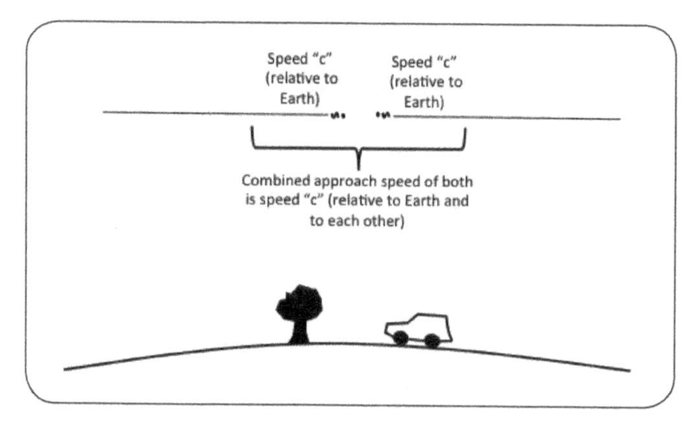

Figure 9-7. Adding velocities of photons considering the effects of relativity. *Two photons approaching each other at speed "c" relative to a third frame-of-reference, like the Earth, would be going speed "c" relative to Earth and to each other.*

The underlying reason that objects moving relative to each other can never exceed speed "c" when their velocities are added together has more to do with the geometry of space-time than the relative speeds of the objects themselves as measured in three-dimensional space. When trying to combine speeds of objects moving through three-dimensional space, we miss the point that the objects are actually moving through four-dimensional space-time. Special Relativity corrects for this restricted perspective. As described earlier, Special Relativity calculates how much time slows and the amount of space compression that occurs when motion through space is considered separately from integrated space-time. This means that judging the velocities of moving bodies with respect to three-dimensional space alone, is a cumbersome exercise. This is because every observer will measure the speeds from their own frame-of-reference and these measures will have to be corrected for any relative motion between the frames-of-reference, as done in the above examples. If we were able to see and measure motion directly in four-dimensional space-time, it would not be so difficult to understand. Observers, regardless of their frames-of-reference, would agree about the velocities without having to make corrections to account for our own movements and spatial perspectives. However, this is not our reality. Because we see three-dimensional space and time as separate things, we must make the corrections using the correction factor from Special Relativity. When these corrections are made, it is not possible to exceed speed "c" in three-dimensional space. The difficulties become even greater when accelerations or gravitational fields are present and the more complex formulations of General Relativity are applicable.

Another explanation that can help us appreciate why it is not possible to go faster than speed "c" is the fundamental relationship between space and time. That relationship is, in fact, speed "c." Space-time combines space (in this case, measured in miles) together with time (in this case, measured in seconds) in the ratio of 186,282 miles per second, or speed "c." Put another way, space and time are combined in the ratio of 186,282 miles of space for every second of time. Attempts to exceed speed "c" would violate the very structure and nature of space-time.

The relationship between high-speed relative motion and the geometry of space-time is difficult to visualize in just three-dimensional space, especially when drawn on a two-dimensional piece of paper, like in this book. However, ignoring this difficulty, and at the risk of even further oversimplifying, let's revisit the figure used earlier showing photons reaching Earth from a distant star. This time, light-clocks are superimposed for the traveling photon's frame-of-reference (as seen from Earth) to help us visualize how time stops and space gets completely compressed to zero. This is shown in Figure 9-8. In this figure, once again, time is represented by the motion of photons through the body of a light-clock. But the photon is traveling at speed "c" towards the Earth, so no vertical motion through the light clock happens, and therefore no time passes. The light clock does not tick off any time. It does not complete any cycles. There is no movement through the body of the light clock. And since no time passes this means that the photons arrive at the Earth instantaneously! Much more on this in Chapter 10.

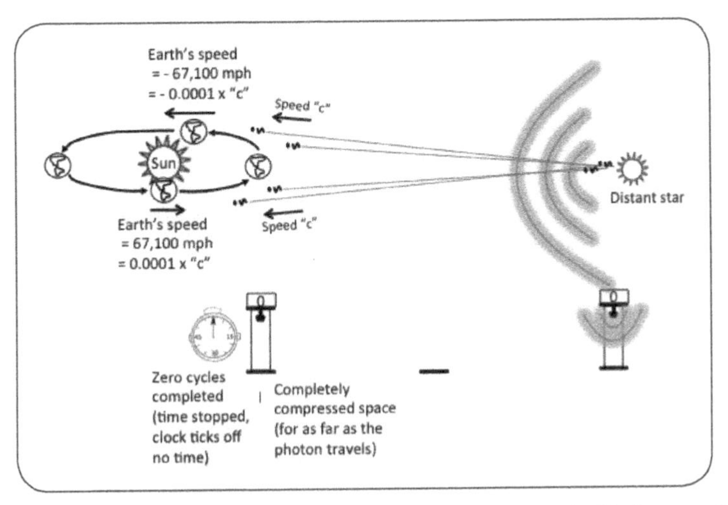

Figure 9-8. Time stops and space compresses to zero at the speed of light. *For photons, always moving at speed "c" relative to all frames-of-reference, space is always compressed to zero and no time passes during the photon's travel. Since no time passes, there is no vertical movement in the light-clock and no time cycles are completed as shown here.*

It is true, of course, as we saw earlier, that the speed we measure for light from a distant star is always speed "c," regardless of Earth's travel direction with respect to the star. And since light always has a measured velocity of speed "c" for all receiving frames-of-reference, it is always associated with complete time slowing and space compression, regardless of travel direction and speed of the receiving body. That is, according to the laws and understanding provided by relativity, photons are everywhere immediately and simultaneously! We will see later in Chapter 10 that there are alternative and equally correct ways of understanding the behavior of photons based on their "quantum" characteristics.

Since, according to relativity, space compresses completely and time stops altogether when speed "c" is reached, intuitively, just based on this conceptual understanding of Einstein's theories, it should not be possible to exceed speed "c." How can space be compressed beyond 100%? And what about all the difficult complications of going backwards in time and perhaps meeting or killing one of your ancestors? Beyond the relativity concepts discussed here, Einstein's theories make these points mathematically. Again, Appendix B provides simple calculations for the examples in this section. And in Chapter 10, the question of whether it's true that there isn't anything that can go faster than speed "c" is discussed. It turns out there are a few exceptions (sort of, maybe).

One last point that is important to make here: while at speed "c," photons are everywhere instantaneously and simultaneously in their frame-of-reference, this is not the case for our frame-of-reference. When we see light (or any electromagnetic radiation) from anywhere in the Universe, we are seeing light arriving from the past, having taken time to get here. The photons themselves may be associated with complete time slowing and space compression, or may exist in some "sweet spot" between time and space. When they arrive in our frame-of-reference, they exhibit the time and space relationships between the Earth and the stars and galaxies they came from. That is, light we see from distant galaxies reflects the space-time separations between those galaxies and us. While the photons may experience complete time slowing (often referred to as time dilation) and complete space compression in their frame-of-reference, they carry information reflecting the separations in time and space based on the frames-of-reference of the emitting and receiving objects. For example, the light we see from the

Andromeda Galaxy is light that was emitted from Andromeda 2 million years ago. The light we see from the Sun is light that was emitted from the Sun 8 minutes ago. More on what this says about the Universe in the context of the even stranger quantum nature of photons (and all subatomic particles) in Chapter 10.

9.3 Simultaneous events and causality. Since time is perceived to move at different rates depending on any relative motions involved, it is not possible to say that simultaneous events in one frame-of-reference will be simultaneous in another frame-of-reference. In fact, sometimes the order of events can even change. However, it is not possible to change the order of events that are causally linked (one causes the other).

In his 1916 book on relativity, Einstein explored whether events that occur simultaneously in one frame-of-reference will also be simultaneous in a second frame-of-reference that is moving relative to the first. His conclusion was an emphatic no; they will not be simultaneous because of the effects of Special Relativity. He demonstrated this with a simple illustration.

He described a train moving along a track with one observer on the train and a second observer beside the train on the ground. Both observers are exactly midway between two bolts of lightening when they simultaneously strike the ground (in the Earth's frame-of-reference). The question posed is this: if the train was moving at near speed "c" in a direction directly away from one lightening bolt and toward the other, would the lightening bolts be seen simultaneously by both observers? The answer is no. This is shown in Figure 9-9.

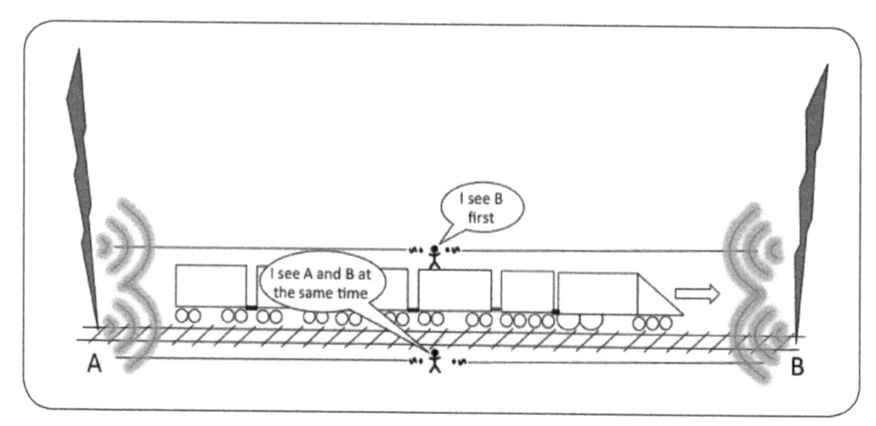

Figure 9-9. Why simultaneous events aren't simultaneous on moving platforms. *Two lightening bolts strike the tracks simultaneously an equal distance from an observer on the ground. A second observer is at that same location along the tracks, but on a train moving at near speed "c". Since the observer on the train is moving toward the lightening bolt at Location B, this lightening bolt is seen first. This is because the train and light from Lightening Bolt B are moving toward each other and meet in the middle. By comparison, the observer on the ground that is stationary with respect to the lightening bolts sees them occurring simultaneously.*

Both observers will not see the lightening bolts simultaneously; the observer on the train sees the lightening bolt from the right (from Location B) as occurring before the lightening bolt from the left (from Location A). But the observer on the ground sees the lightning bolts simultaneously. The reason the observer on the train sees the lightning from location B first is because the train is moving toward the lightning at location B at near speed "c," so the distance traveled by the light is less (as the train and light meet in the middle). Similarly, the light from Location A has further to travel to "catch-up" to the moving train, and thus takes longer to get to the train-bound observer. The observer beside the train (on the ground) sees both lightening bolts at the same time since this observer is not moving relative to either lightning bolt; the light has the same distance to travel in both cases. This is another way of looking at how Special Relativity works. The point is that simultaneous events in one frame-of-reference will not be simultaneous in another frame-of-reference. As we will see shortly, this lack of simultaneity is directly related to the slowing of time and compressing of space.

In addition to making the idea of simultaneity meaningless for observers in relative movement, relative movement can even change the order in which events occur. This is easy to illustrate by simply expanding the train example shown in Figure 9-9. The expansion is shown in Figure 9-10 in which we also consider a third frame-of-reference, that of a train on a parallel set of tracks traveling in the opposite direction to the first train. In this case, the observer moving toward the lightening bolt at Location B will see the lightening at Location B before seeing the lightening at Location A. Alternatively, the observer moving toward the lightening bolt at Location A will see Lightening Bolt A as occurring first. Thus, the relative motion of observers changes the order in which these events are perceived to occur.

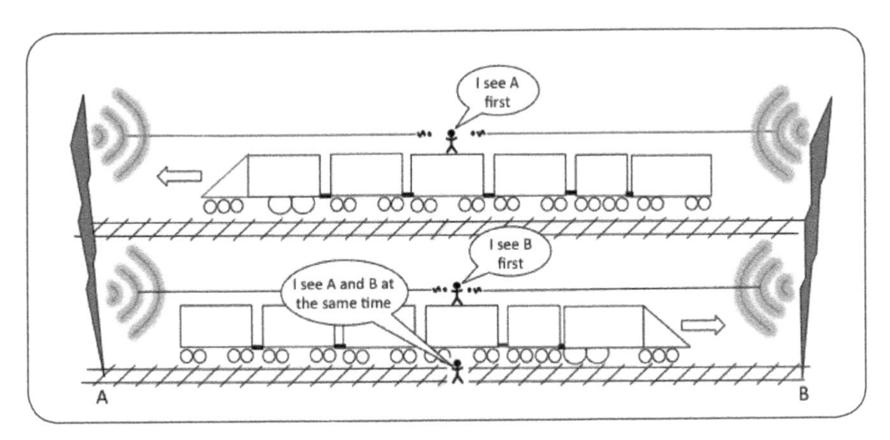

Figure 9-10. Relative motion can change the order that events happen. *Two lightening bolts strike the tracks simultaneously an equal distance from an observer on the ground next to train tracks. This observer sees the lightening bolts as occurring simultaneously. However, observers on two moving trains, each moving in the opposite direction, see the lightening bolts as occurring in the opposite order. The observer moving toward Lightening Bolt A, sees lightening A as occurring first, while the observer moving toward Lightening Bolt B, sees lightening B first.*

But can relative motion change the order of events such that results happen before causes? For example, if a man were to fire a gun at another person, killing that person, is it possible for an observer on a moving platform to see the victim die before the gun was fired? It turns out that this is ***not*** possible, according to Einstein, because this would require the bullet to travel faster than speed "c," which we know is not possible. This situation is illustrated in Figure 9-11 and described next.

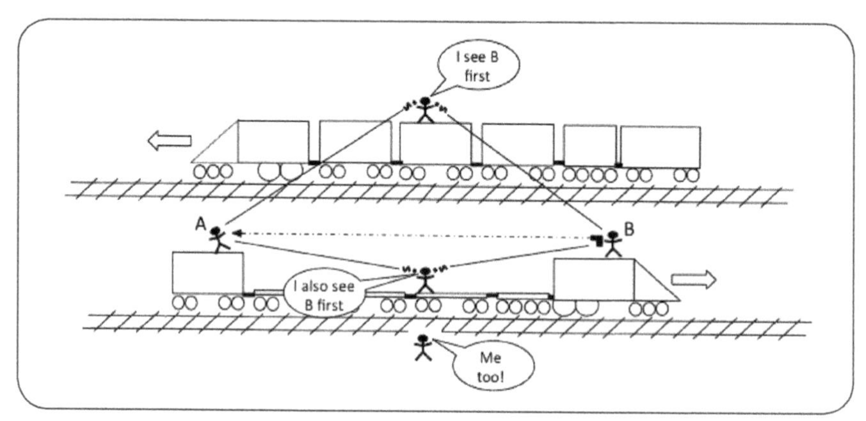

Figure 9-11. Relative motion cannot change the order of events when they are causally related. *Two observers watch a person at Location B fire a gun, killing a victim at Location A. Even for observers on trains moving at near speed "c" in opposite directions, the firing of the gun will always be seen as occurring before the victim gets killed.*

In this case, two observers witness the events of a person firing a gun from Location B and killing a victim at Location A. One observer is on the same train (same frame-of-reference) as both the killer and the victim. A second observer is on a train traveling in the opposite direction. Both trains are traveling at near speed "c" relative to each other. (As described above, even if the two trains are traveling at near speed "c" relative to the ground, they will still not be traveling faster than speed "c" relative to each other.)

In this example, the observer that is located on the same train as the killer and victim sees the killer fire the gun before the victim is killed. The reasons for this are obvious; it takes some time for the bullet to travel from the gun to the victim. Since the observer, in this case, is in the same frame-of-reference as the killer and victim, there is no time slowing or space compression involved. The observer sees the victim getting killed after the killer fires the gun.

But what about an observer located on the train that is traveling in the opposite direction? In this case, the moving observer is traveling toward the event of the victim being killed and away from the event of the killer firing the gun. In both cases, the relative speed is near speed "c." Therefore, this "moving" observer will see the event of the victim being killed as occurring before the "non-moving" or "stationary" observer on the same train as the killer and victim sees it. Similarly, the "moving" observer will see the killer firing the gun as occurring later than the "stationary" observer on the same train as the killer and victim. But can these time alterations be great enough to change the order in which the two events are seen to occur — as it did in the lightening example? Can the "moving" observer see the victim die before the killer fires the gun? The answer is no. The ordering of the two events (firing the gun and killing the victim) will always occur in the same sequence. This is because the two events are causally linked, in this case, by the passing of a bullet between the killer and the victim, a bullet that will always travel slower than the speed of light as described further below.

So here's the reason that the relative motion cannot change the order such that the victim dies before the gun is fired. Since the bullet will always travel more slowly than the speed of light, the time separation between the two events will be longer than the time it takes for light to travel between the two locations, regardless of any relative movement of the observers. This means that no matter how fast an observer moves relative to the events, the ordering of the events will always be the same. Even if the bullet were to travel at the speed of light (a laser bullet), there would still be a definitive time separation between the two events (the time it takes for light to move between the two). The "moving" observer could see the two events as occurring very, very, very close in time, but always in the same sequence. It would be necessary for the laser bullet, or the light associated with seeing the events, to exceed speed "c" in order to change the event sequence; and we know this is not possible. Only events that are not linked causally (i.e., requiring some information sharing between the two) can have their sequences changed through relative movements (such as the lightening bolts that have no causal association).

9.4 The relationship between event ordering, the slowing of time and the compression of space. The effects of relative motion on the sequencing of events are directly related to the associated slowing of time and compression of space. This section describes this relationship.

Let's examine the moving trains example further to understand why changes in event timing are related to the slowing of time and compression of space; and why time slowing and space compression occur. Again, we have two trains moving at near the speed of light past each other in opposite directions on parallel tracks. We look at two events relative to the passing trains as experienced by the observers on the trains. This time the two events are: (A) the event when the front of Train 1 and the back of Train 2 are at the same location along the tracks; and (B) the event when the front of Train 2 and the back of Train 1 are lined up. This is shown in Figure 9-12. We will look at these events from the perspective of each train, plus the observer on the ground.

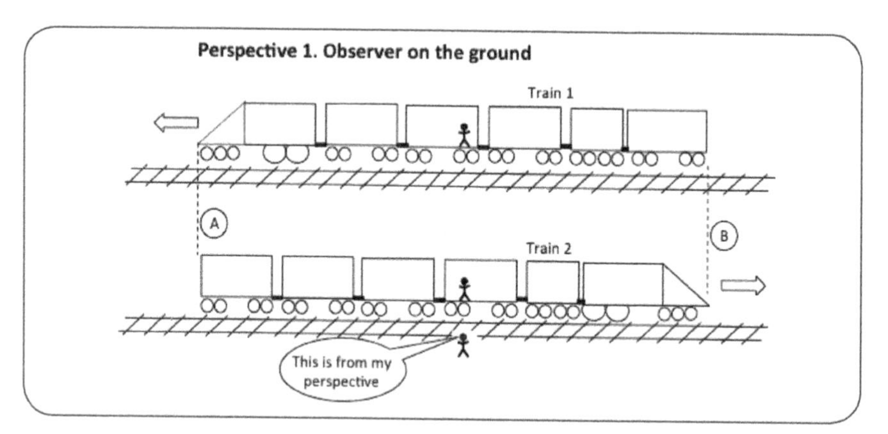

Figure 9-12. Passing trains define events for analyzing the effects of Special Relativity. *Passing trains provide reference events that allow us to see the relationship between relative motion and event ordering. It shows us how this is due to time slowing and space compression.*

Perspective 1. Observer on the ground. Events A and B occur at the same time because the trains are the same length and moving at the same relative speed, although in opposite directions.

Perspective 2. Observer on Train 1. Event A occurs before Event B. Again, this is because the light (information) emanating from Event A and the observer on Train 1 are moving toward each other so that the light, which always travels at speed "c," has less distance to travel. Also, the observer on Train 1 is moving away from light (information) coming from Event B, making the distance the light travels longer. For this reason, Event B is seen by the observer on Train 1 to occur after Event A. This means that when Event A occurs for the observer on Train 1, Event B has yet to occur. Similarly, when Event B happens, Event A has already occurred. This corresponds to the apparent compression of space on Train 2 as seen by the observer on Train 1. This is shown in Figures 9-13 (showing Event A) and 9-14 (showing Event B).

And this makes sense given what we know about the constancy of the speed of light, regardless of relative motions, and what this means for the relative passage of time on the respective moving platforms. Space is compressed and time slowed on the moving train as seen from the reference train (i.e., the frame-of-reference selected to be "stationary"). Examination of the two figures makes this graphically obvious. For the case of Train 2 as seen by the observer on Train 1, Event A happens before Event B and Train 2 takes some time to get from Event A to Event B. That is, time moves more slowly on Train 2 as seen by the observer on Train 1, and this relates directly to space compression on Train 2.

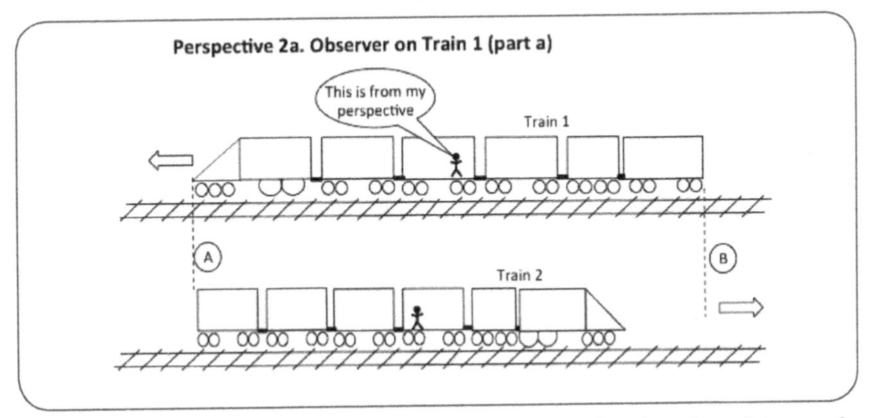

Figure 9-13. Relative motion changes the order of events and makes slowed time and compressed space obvious (Train 1, Part A). *Passing trains illustrate why time slows and space contracts due to Special Relativity. For the observer on Train 1, Event A occurs before Event B. Graphically, this shows why space is compressed and time slowed on Train 2 as seen by the observer on Train 1.*

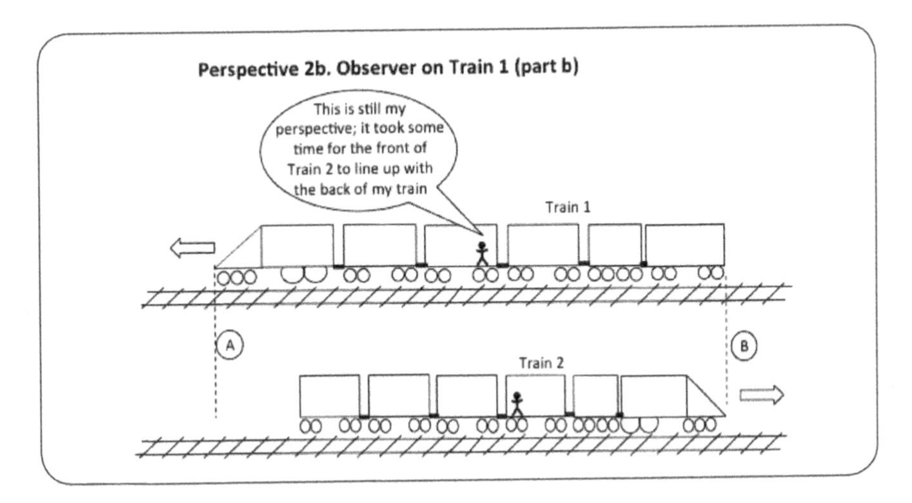

Figure 9-14. Relative motion changes the order of events and makes slowed time and compressed space obvious (Train 1, Part B). *For the observer on Train 1, Event A occurs before Event B. This means that when Event B happens, Event A has already occurred. This further illustrates why space is compressed and time is slowed on Train 2 as seen by the observer on Train 1.*

Perspective 3. Observer on Train 2. Of course, things are reversed from the perspective of Train 2. The observer on Train 2 sees Event B as happening first followed by Event A. This means that from Train 2's perspective, time is slowed and space compressed on Train 1. This is shown in Figures 9-15 and 9-16, respectively.

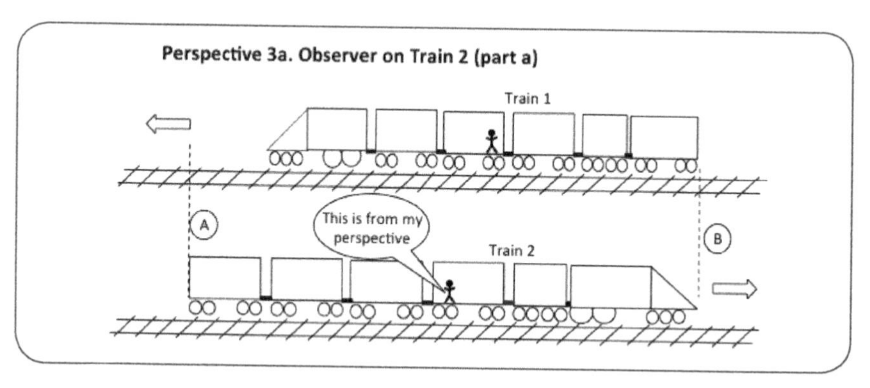

Figure 9-15. Relative motion changes the order of events and makes slowed time and compressed space obvious (Train 2, Part A). *For the observer on Train 2, Event B occurs before Event A. This means that when Event B happens, Event A hasn't yet occurred. This graphically illustrates why space is compressed and time slowed on Train 1 as seen from Train 2.*

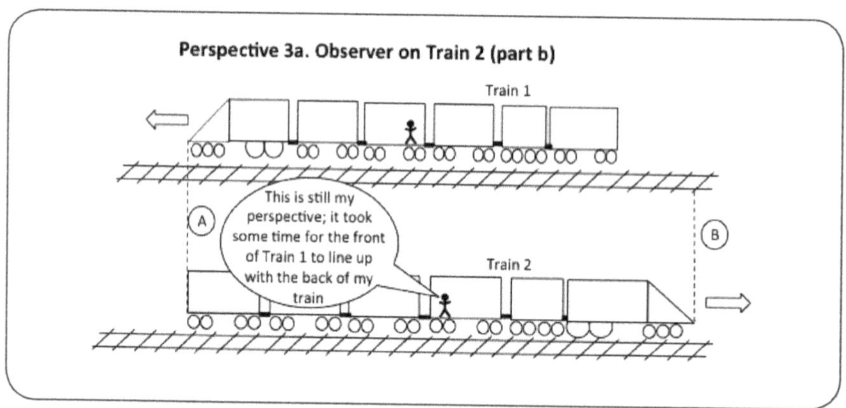

Figure 9-16. Relative motion changes the order of events and makes slowed time and compressed space obvious (Train 2, Part B). *For the observer on Train 2, Event B occurs before Event A. This means that when Event A happens, Event B has already occurred. This further illustrates why space is compressed and time is slowed on Train 1 as judged from Train 2.*

In this example, the ordering of Events A and B are reversed depending on the observer's frame-of-reference. The observer on Train 1 sees A occurring first and then B; and sees Train 2 as shorter (Figures 9-13 and 9-14). The observer on Train 2 sees B occurring first and Train 1 as shorter (Figures 9-15 and 9-16). Both are right due to the effects of Special Relativity. Time is slower and space more compressed on the moving train as seen by the observer on the "non-moving" train. And it depends on which train is selected as the observing ("stationary") frame-of-reference.

Chapter 10

What Relativity Says About the Universe: Part 2 — How the Universe is Made

This chapter addresses the following topics and questions:

10.1 Moving through space-time. *Given the one-directional nature of time and its integration with the three dimensions of space, how is it that we move through space-time? How does the universal speed limit, "c," restrict our movement through space-time?*

10.2 Relativity limits information access. *How does relativity limit what we can know?*

10.3 Time it takes for light to travel across space. *Is it really true that it takes light many years to reach us from distant stars and galaxies? Well no, but why not?*

10.4 Exceeding the universal speed limit. *Really, isn't there anything that can go faster than speed "c"? Yes, there may be, but what is it?*

10.5 The appearance of compressed space or actual compressed space? *Do things actually become shorter with the compression of space or is it just the appearance of things (and space) shortening? Yes. And you will see why, too.*

Continue on page 130

Continue from previous page

10.6 How objects bend space-time. *How do objects tell space-time to bend? How does space-time know there is a large massive body nearby and that it needs to bend?*

10.7 The structure of space-time. *Is space-time really empty? What is space-time made of? What is space-time, really?*

10.8 $E=mc^2$. *What does $E=mc^2$ have to do with relativity anyway? How are energy and mass interchangeable? Can energy really turn into mass and vise-versa? What does this say about how the Universe is put together?*

10.9 Limits of relativity's scope and competence. *Are there limits to what relativity can tell us about the Universe? What are these limits and how can we fill-in these gaps?*

Einstein's theories have profound implications about how we understand the Universe. This chapter continues the discussion of important areas where our understanding of the Universe has been shaped and changed by Einstein's theories of relativity.

10.1 Moving through space-time. This section describes how time is integrated with the three space dimensions. It also explores how the one-directional time dimension and universal speed limit "c" restricts our movement through space-time?

As described throughout this book, our movement through space-time affects our perception of space and time when considered separately — *"moving clocks run slow"* and all that. In this section the integrated space-time is represented graphically to show how the integration of space with time that moves from past to future defines and constrains our movements in space-time. It shows how movements through space are associated with changes in time, and shows how our movements through space-time are restricted by: (1) the present-only, forward-moving nature of time and (2) the universal speed limit, speed "c."

To do this, it is common to depict space-time using a two-dimensional graph with time measured along the vertical axis and the three space dimensions combined into a single spatial component plotted along the horizontal axis. This graph, called a space-time chart, is shown in Figure 10-1. It visually illustrates some key features of Einstein's theories. First, it puts time and space on one graph to emphasize that the two exist and work together. It graphically shows that locations in space-time are defined as points that include both spatial and time components. Second, it shows that time values can be in the past, present or future. Locations of past events are shown below the horizontal axis while future events appear above the horizontal axis. Events happening right now appear along the horizontal axis. They will appear below the horizontal axis the next time the graph is drawn, reflecting the one-way, forward-moving nature of time, always moving from present to future. It is typical to show our current location, or the location of an object or event of particular interest, at the center of the graph.

Events A through E in the figure show example events illustrating how the chart works. Event A is happening right now. Since Event A is at the center of the chart, it is our current location or the location of an event of interest. Event B is also happening right now at a distance of 2 light-years away from Event A. Because the chart combines all three spatial dimensions along the single horizontal axis, it is only possible to know that Event B is two light-years away from Event A along an unspecified combination of the three spatial dimensions. Event C happened at a spatial distance of 1 light-year away from Event A and 1.5 years in the past. Event D will happen in the future, one year from now at a distance of 1 light-year away from Event A. And Event E will happen in 2 years at a distance of 1 light-year away from Event A.

Finally, this graph depicts a single frame-of-reference against which locations and movements through space-time can be shown. Objects remaining at the same spatial location on this chart over time would be shown at the same position along the horizontal axis, but at different locations vertically along the vertical (time) axis reflecting the movement through time from past to present to future. However, remaining in one location along the spatial (horizontal) dimension is defined as stationary only for the frame-of-reference depicted by the chart. It might be seen as being in motion (changing position along the horizontal axis) as judged from another frame-of-reference. If that other frame-of-reference were shown on another space-time chart, the movement would be represented as changed positions along the horizontal (space) axis. It depends on the selected ("stationary") frame-of-reference depicted on the chart.

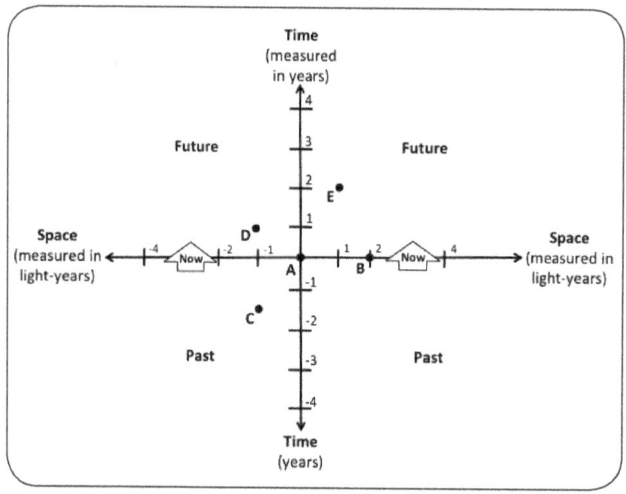

Figure 10-1. Locations are defined by a spatial coordinate and a time coordinate. *Time and space exist together, so locations and events must have a spatial component and a time component. This is shown here on a space-time diagram. We can control our movements through space, but not time.*

Space-time charts are especially useful for defining limits to our movements, both past and future. Since it is not possible to travel faster than speed "c," there are areas of space-time that we can and cannot travel to. Figure 10-2 shows how our movements are restricted by this limitation. It assumes, as is customary, that our current location is at the center of the chart where the horizontal (space) and vertical (time) axes cross. The shaded area in the upper part of the chart shows parts of space-time that we ***can*** possibly travel to (without exceeding speed "c"). The shaded area in the lower portion of the chart shows space-time locations where we ***could*** have been in the past (again without having traveled faster than speed "c"). Travel along the diagonal sides of the shaded areas would require traveling at the speed of light (one light-year in one year). Further, even within the shaded areas, it is not possible to exceed the speed of light; that is, traveling faster than one light-year per year. Lines representing the speed of light within the shaded areas are also shown in the figure, indicating examples of travel limits within space-time. The darkly shaded area associated with the example future location indicates travel limits from that future location in space-time.

The areas of the chart that are ***not*** shaded are areas of space-time that we ***cannot*** experience (e.g., future locations we cannot reach and past locations we could not have been in the past, judging from our current location). These areas of space-time exist, but would require exceeding speed "c" for us to experience.

The curvy black line in Figure 10-2 extending to and from the center of the chart (current location in space-time) indicates an actual travel path through space-time. This is called a "worldline." It represents an actual path of travel through space-time for a given person or object. Depending on the speed of travel, the slope and shape of worldlines will vary. The only limitation is that the movement cannot exceed speed "c." That is to say, it cannot leave the shaded area of the chart or exceed the slope of the diagonals defining the edge of the shaded area. This means we can't move more than one light-year per year (e.g., it is not possible to move more than one light-year along the horizontal axis within one year along the vertical axis). This would require exceeding speed "c" in which one light-year is traversed in one year.

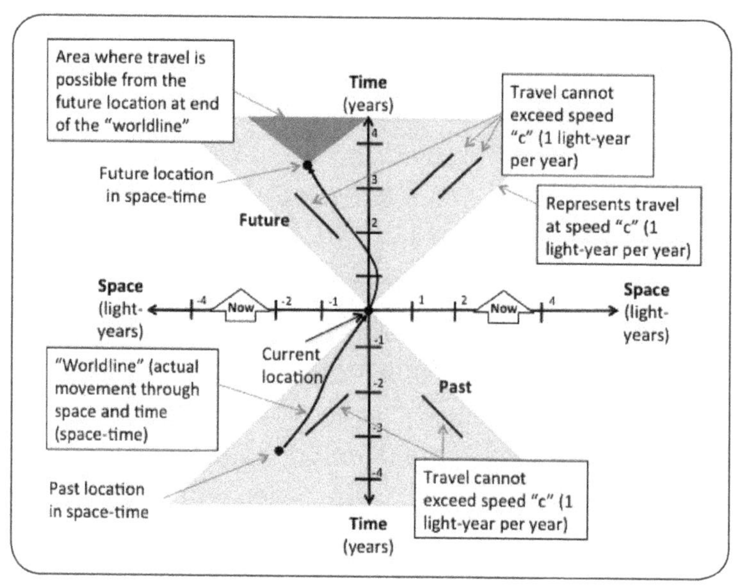

Figure 10-2. Space-time diagram with example of possible travel. *The black line connecting a past location with a future location is called a "worldline" and represents one possible path through space-time. The diagonal lines represent traveling at speed "c." The shaded areas indicate areas where past travel was possible, or where future travel is possible, based on the current location at the center.*

If we were to draw worldlines for the twins described in the previous section on a space-time diagram, the two worldlines (for the two twins) drawn from the Earth's frame-of-reference would look like those shown in Figure 10-3. The worldline for the twin that traveled to the distant star at 80% of the speed of light and then returned to Earth is shown to the right of the vertical (time) axis. You will recall that the twin that traveled to the distant star was the one who experienced slowed time and compressed space and rejoined the Earth-bound twin much younger than the twin who remained on Earth. Worldlines that are longer on these space-time diagrams are associated with relative movements characterized by compressed space and slowed time (for the frame-of-reference being shown). Also, worldlines that deviate from straight lines indicate changes in speed or direction during the trip, like when the traveling twin reached the distant star and turned back to Earth. Because non-straight worldlines are associated with changes in speed or direction, they also indicate that accelerations must have occurred; accelerations that add additional time slowing beyond that just associated with the longer worldline itself.

Accelerations are also associated with changes in frame-of-reference. For example, when the traveling twin rejoined her stay-at-home twin on

Earth, the traveling twin's worldline changed direction on the chart and began moving vertically along the time axis. This action of rejoining the stay-at-home twin's frame-of-reference also had the effect of "locking-in" the slowed time that the traveling twin experienced relative to Earth's frame-of-reference during the trip to the distant star and back. This means that after the twins were both back together in Earth's frame-of-reference, the traveling twin would be more than 20 years younger than her stay-at-home twin, as noted earlier. This age difference is due to all of the relative motion and accelerations the traveling twin experienced relative to Earth's frame-of-reference.

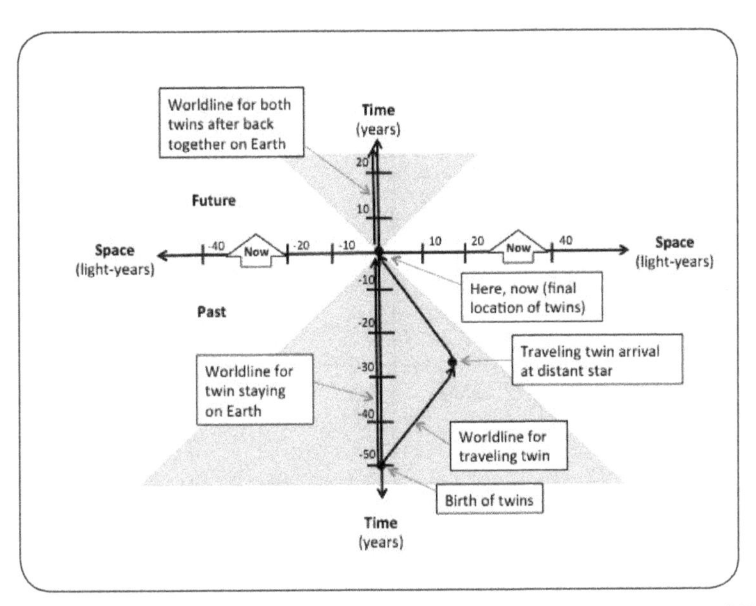

Figure 10-3. Space-time diagram for the twin paradox. *This figure compares worldlines for Einstein's hypothetical twins described earlier. The worldline to the right of the vertical axis is associated with the twin that traveled to a distant star and then returned to Earth at 80% of speed "c." The worldline to the left and parallel with the time axis is associated with the twin that stayed at home and for both twins after the traveling twin returned to Earth.*

10.2 Relativity limits information access. In addition to limiting where we can go, the structure of space-time limits what we can know. This section explains.

The structure of space-time does more than set limits on the movements of objects and people. It also sets limits on the communication of information. It defines areas of space-time we can receive information from or send information to. This is because the same speed limit (speed "c") for traveling through space-time also applies to communication of information through space-time. Like people and things, information cannot exceed speed "c".

There are several important ramifications that derive from these limitations on sending and receiving information that are illustrated by these space-time charts. These are: (1) there are areas of the space-time from which we cannot receive information and to which we cannot send information (i.e., outside the shaded areas); (2) information that we receive always reflects events that have happened in the past (below the horizontal axis); and (3) we can only send information to the future (above the horizontal axis). Figure 10-4 shows these ramifications graphically using the space-time chart.

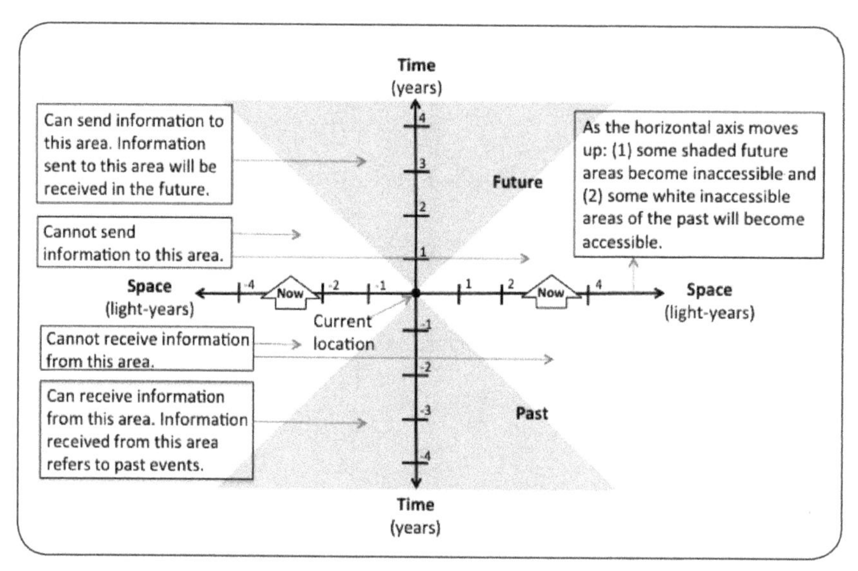

Figure 10-4. Space-time diagrams establish limits on information communication. *Just as space-time limits where we can go, it also sets limits on the communication of information.*

The fastest mechanism for communication across space-time is with light (or other forms of electromagnetic radiation, such as radio waves) because these communication modes travel at speed "c." The diagonal lines on the space-time charts represent the fastest communication possible between the current location (shown at the center of the chart) and distant locations of space-time. For example, information about an event that occurred at a location one light-year away can reach us in one year at the speed of light. This information will reflect events and actions that occurred one year ago.

An illustration of how this works in the real Universe is for communications that take place between Earth and the latest Mars exploration vehicle, Curiosity. Depending on where Earth and Mars are in their respective orbits, it takes between 4 and 24 minutes for radio signals to get to Mars from Earth. On average, Mars is 14 light-minutes from Earth. Therefore, it takes an average of 14 minutes to receive information from Mars and another 14 minutes to send information to Mars. It is therefore not possible to know what Curiosity is doing at this instant and it is not possible to instruct Curiosity to take desired actions immediately. This is an important issue for the Mars Curiosity mission. Instructions sent to the Curiosity robot on Mars (e.g., to tell it to stop or change direction) are made based on information about the Curiosity's location and direction of movement reflecting the situation an average of 14 minutes earlier. And Curiosity receives these instructions an average of 14 minutes later, a total time lag of 28 minutes. This is shown in Figure 10-5. Note that the figure shows space in light-minutes and time in minutes. Speed "c" in light-minutes is one light-minute per minute.

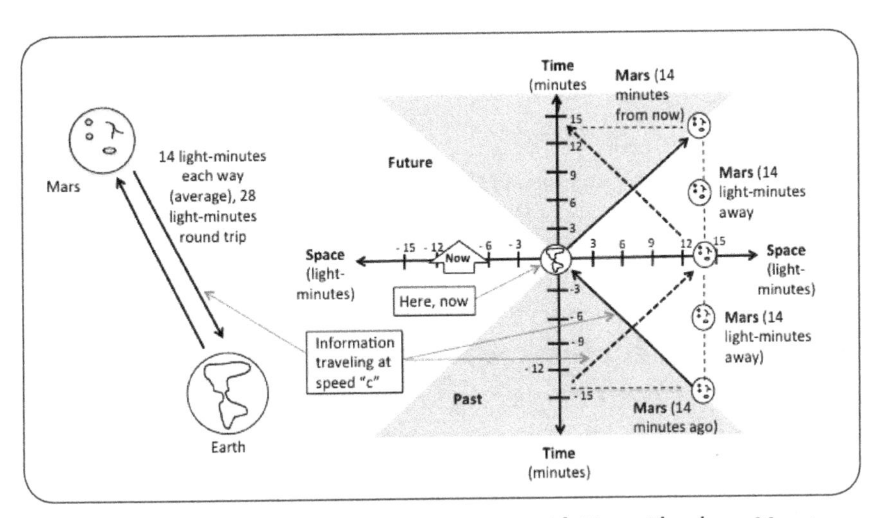

Figure 10-5. Space-time diagram for communication with Mars. *The planet Mars is an average of 14 light-minutes from Earth. This means that communication between Earth and Mars takes an average of 14 minutes each way.*

This shows that space-time areas in the un-shaded portions of space-time diagrams do exist, just that we cannot visit them and cannot communicate with them. In fact, we can visit and communicate with these areas of ***space***, just at a later ***time***. However, since locations are defined by both a position in space and a position in time, it is the un-shaded areas of integrated space-time that we cannot reach or communicate with. Considering space and time separately, we can visit and communicate with all locations in space and do experience all moments of time that occur within the short duration of our lives. It's just that some locations in the integrated space-time cannot be reached or communicated with (from our current location).

Light from distant galaxies takes even longer to reach us. For example, it takes light 2 million years to reach us from the Andromeda Galaxy, the closest galaxy to our own. We can only know about things that happened in Andromeda 2 million years ago. To communicate across 2 million light-years takes 2 million years time; assuming the signal travels at speed "c." This is shown in figure 10-6.

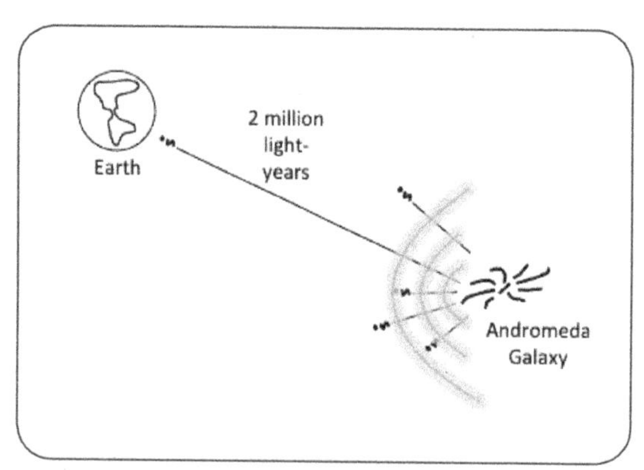

Figure 10-6. Distance to the Andromeda Galaxy. *The galaxy Andromeda is 2 million light-years from Earth. This means that we can only find out about things that happened in Andromeda 2 million years ago.*

10.3 Time it takes for light to travel across space. It is routine to describe distances between stars and galaxies in terms of light years, and to assume that it takes light that many years to get here across these distances as discussed in this book. However, it is not entirely accurate to say that it takes light many years to reach us from distant stars and galaxies.

As discussed earlier, travel at speeds approaching the speed of light results in slowed time and compressed space when judged from frames-of-reference chosen to be stationary. We saw this illustrated, for example, in Einstein's twin paradox where the traveling twin aged less and experienced compressed space compared to her non-traveling twin, who remained on Earth. Also noted earlier, travel at speed "c" results in total space compression and complete time slowing. Does this mean that photons experience no time passage and complete space compression? For example, do photons get here immediately from say, the Andromeda galaxy? Do they get everywhere immediately, since light travels at speed "c" relative to ***all*** frames-of-reference? Relativity would seem to say yes.

However, the answers to these questions are more complicated than they seem. From a purely relativistic perspective, it would hold that photons do experience complete space compression and time slowing relative to all frames-of-reference. But this simple explanation ignores the fact that photons are "quantum particles" that are also subject to the laws of quantum mechanics, which state that it is not possible to measure ***both*** a quantum particle's speed ***and*** location, at least not with complete certainty. The strange nature and behavior of quantum particles is beyond the scope of this book. An excellent discussion of the quantum nature and behavior of light is provided in Richard Feynman's book, *QED: the Strange Theory of Light and Matter.* It is enough to say here that the laws of quantum mechanics and those of relativity are not entirely compatible with each other despite attempts by Einstein and other scientists over many years to combine them. More on the strange behavior of quantum particles is presented later in this chapter.

In spite of the quantum mechanical difficulty of simultaneously measuring both the speed and location of quantum particles (including photons), what we can say from just the perspective of relativity is that photons appear to travel at speed "c" relative to all frames-of-reference. Therefore, just considering the effects of relativity (and ignoring the complications of quantum mechanics) it appears that photons do reach everything and everywhere in no time at all with the intervening space compressed to zero.

But while photons may experience completely compressed space and no time passage the distances they traverse are real, and the information carried by photons reflects the time-space separations and the relative motions of the frames-of-references being connected. For example, appearance of the Andromeda galaxy reflects how Andromeda appeared 2 millions years ago. This is because Earth and Andromeda are separated by 2 million light years, and for the purposes of this example, assumed to not be moving relative to each other, certainly not at speeds approaching speed "c." This is shown in Figure 10-7. Like many figures in this book, Figure 10-7 is an oversimplification drawn on a flat two-dimensional page, but it helps us imagine and visualize the phenomenon that plays out in four-dimensional space-time.

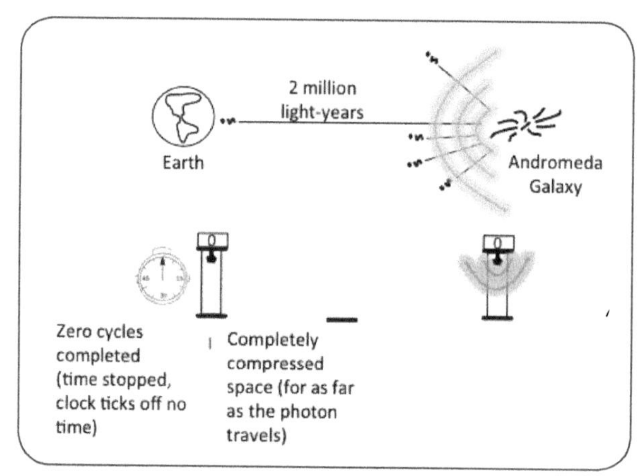

Figure 10-7. Photons experience complete time slowing and space compression. *Photons moving at speed "c" from Andromeda to Earth experience total space compression and complete time slowing due to the Laws of Relativity. At speed "c," the photons can be thought of as experiencing no vertical movement through the light clock, and thus no time ticks off. They arrive at Earth immediately, even though the information they carry is 2 million years old.*

Even though relativity suggests that photons may experience complete space compression and time stopping, the time-space distances covered are real for the photons too. They may arrive everywhere immediately, but even at the speed of light with complete space compression and no time passage, the photons are affected by and experience the intervening space. For example, they respond to any gravitational influences created by massive objects that may be located between the light source and destination, and of course, solid objects they may strike will block their passage. Even though from the photon's perspective, space may be compressed to zero, they bend around stars and galaxies along their route where space-time is bent by gravity. While the intervening space may be compressed to zero, the photons must still go through it. Go figure.

10.4 Exceeding the universal speed limit. It is not completely accurate to say that there isn't anything that can go faster than speed "c." This section explains.

So isn't there anything that can go faster that speed "c"? Well, it turns out there are a few of things that seem to do this, sort of. These are described here. Also, there is evidence that immediately after the Big Bang that created the Universe, the newly created Universe itself, along with all the things in it, expanded at a rate that was faster than speed "c." This was perhaps before the current laws of physics were created or began to operate. For the Universe as it is today, there are three situations where exceeding the speed of light is at least implied. These are introduced below and then expanded upon in separate sections.

The first is space itself. Because the Universe is expanding, the further away in space you go the faster that portion of space (and the stars and galaxies contained there) is moving away. It is therefore possible to reach a point where space is actually moving away faster than the speed-of-light! Of course, because these remote areas of the Universe are moving away at speeds faster than light, it is not possible to see objects in that part of space. We will see in a minute that it is not really these parts of space that are moving faster than speed "c" as they appear, but just that they are beyond the "visible horizon" of space-time.

Second, some subatomic particles exhibit behaviors that, at least, imply faster-than-light communication. It's not that the particles travel faster than light, just that there appears to be the ability of particles that were once together to maintain knowledge of the other's behavior no matter how far away they become; even when the sharing of this information would seem to require faster-than-light communication. This is called "entanglement" and scientists do not understand how this happens, but have performed experiments that indicate that it does.

Third are theoretical things called wormholes. Wormholes are only hypothesized to exist based on mathematical models. It is thought that wormholes connecting remote areas of space-time could be formed between black holes. These are more like shortcuts connecting remote locations in space that theoretically could allow travel across these distances in less time than the separations measured in light years would otherwise allow.

All three of these situations are described in further detail in this section.

The Universe is expanding. This was the great discovery of astronomer Edwin Hubble in the early 20th century. When Hubble tried to catalog the motions of the many galaxies scattered throughout the visible Universe, he was surprised to find that all distant galaxies were moving away from us, and that the further away the galaxies were located, the faster they were moving away. This discovery is what led Hubble and other scientists to conclude that the entire Universe must be expanding. In addition, they concluded that if the entire Universe is expanding, then at one time in the distant past the entire Universe must have been contained at a single point. The Universe must have emerged from a single point in a giant explosive event. This event is now called the Big Bang. It marks the beginning of the Universe that has been calculated to have occurred about 13.8 billion years ago.

The discovery that the Universe is expanding has immense ramifications, not only about how the Universe started, but also about its structure. As the Universe expands it contains more space. That is, space itself is being created with the expansion. And as more space is created, and the Universe expands, the further reaches of the Universe move away ever faster. A point is reached when an area of space is so far away that it is moving away from us at a speed faster than speed "c." Light from these very distant galaxies cannot reach us because this would require speeds faster than speed "c." These parts of the Universe are therefore not visible to us. The point where space is receding faster than speed "c" represents a horizon in the Universe beyond which we cannot see, somewhat analogous to Earth's horizon. This is shown in Figure 10-8.

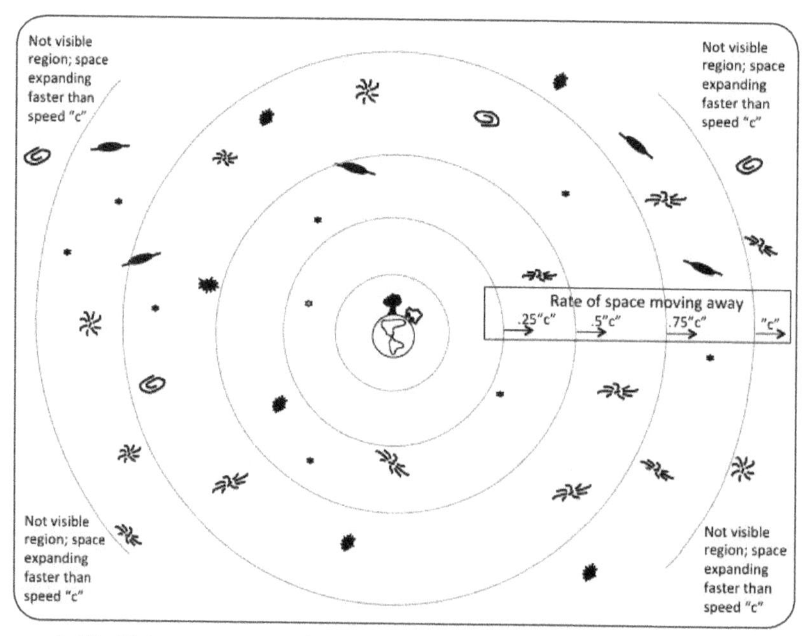

Figure 10-8. The Universe is expanding. *The Universe is expanding, so the further out we look the faster that portion of space is moving away. A point is reached where objects are moving away faster than speed "c." This is a horizon in space beyond which we cannot see.*

But this view of the Universe is not entirely accurate either, because it depicts only space and not space-time. We know that space does not exist independent of time and that it is really space-time that separates the various places in the Universe. The Universe, from a more complete space-time perspective, can be thought of as an expanding sphere (like a balloon being inflated). The two-dimensional surface of the sphere represents space alone without its integrated partner time. The diameter of the sphere represents time. This is shown in Figure 10-9. As the sphere gets bigger (like a balloon being inflated), the objects on the surface (space) appear to be moving further apart. The further away they get, the faster they move, until at some point they are moving faster than speed "c" and disappear. But when the diameter of the sphere (time) is considered, we realize that it is not that more distant objects are moving faster than speed "c," it is just that they have disappeared over the space-time horizon created by the curvature of space-time. It is only when the Universe is recognized as being composed of space-time that we have a more realistic view of how the Universe works. Distant parts of the Universe are not really moving faster than speed "c," they are just disappearing over the space-time horizon.

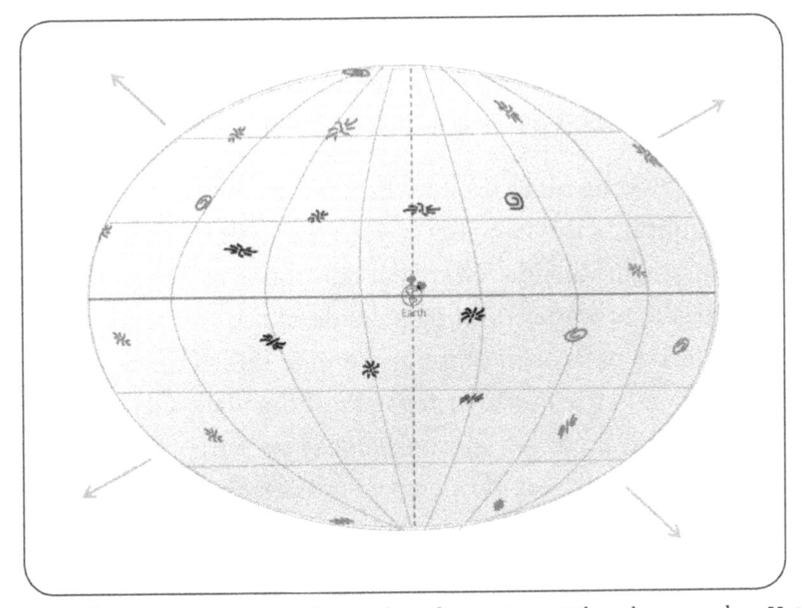

Figure 10-9. The Universe is expanding in four dimensions. *When the expanding Universe is understood as a four-dimensional space-time sphere, we realize that distant objects in space are really just disappearing over the space-time horizon and therefore just not visible to us.*

Entanglement. As introduced earlier, scientists have discovered a very strange phenomenon, called entanglement, in which related characteristics of once-linked subatomic particles, properties such as their "spin" or "polarity," remain correlated no matter how far apart the particles subsequently move. Even more important, the particles seem to retain knowledge of the other's current condition with regard to these characteristics no matter how far apart they move.

Einstein realized that the implications of these conclusions were wide-ranging, and required or at least implied faster-than-light communication. He actively engaged in the debate over the conclusions being drawn, because these implications, if true, would violate the rules of Special Relativity. Einstein argued that there must be an explanation other than faster than light communication.

Entanglement is one of the many strange things that happen at the quantum (subatomic) level. Perhaps the clearest example of this quantum strangeness is related to the behavior of light regarding the way it manifests itself as both a wave and a particle. This book tries to capture this dual nature of light by depicting it as a squiggle (wave) and dot (particle). But it's more than just that light behaves like a wave and like a particle. It is the way it switches between the two behaviors that is so strange. This behavior is illustrated by the famous "double slit" experiment.

In the double slit experiment light is directed toward a board with two slits in it. When the light hits the two slits it goes through them both creating waves of light on the other side. That is, it continues to behave like waves and creates separate sets of waves on the backside of the two slits. The two new sets of waves interact with each other and create interference patterns where the waves cross. When the crests or troughs of the waves line up as they cross, they add their intensities and light becomes either more or less intense in those areas. When the crest of one wave crosses the trough of another wave, they counteract each other and the light intensity goes to zero. When either two crests or two troughs cross, they add their intensities. This results in a pattern of bright and dark stripes on the wall behind the slits.

The question scientists asked was, what happens if you send just one photon at a time through the slits? Would it still create a wave interference pattern, or would the single photon go through just one or the other slit, and not create wave interference patterns behind the board? As shown on the top portion of Figure 10-10, the wave interference patterns are still created when multiple photons are sent toward the slits individually. That is, the photons still behave like waves even when sent just one at a time. But why didn't the photon go through just one of the slits? Why did it still make a wave interference pattern?

To answer this question, scientists placed a sensor at the slits to see which one the photon went through. When this was done, the photons did go through just one or the other slit, and behaved like particles, not waves. After repeating this many times, a pattern of two lines aligning with the slits was created on the back wall. No wave interference pattern was created. This is shown on the bottom portion of Figure 10-10. It appears that the act of monitoring the photons, the act of watching or observing, changes the way photons behave. And it's not just for photons. This effect has also been demonstrated for other subatomic particles, and even for small molecules, too. For example, when electrons were sent through the slits, the same effects were found. Did the photons and electrons know they were being watched? It appears so.[1]

1 This is difficult to visualize and understand from a textual description, even with a diagram like Figure 10-10. There are a couple of short videos on YouTube that offer a nice visual demonstration: www.youtube.com/watch?v=DfPeprQ7oGc and www.youtube.com/watch?v=DsxA7OU7fRO.

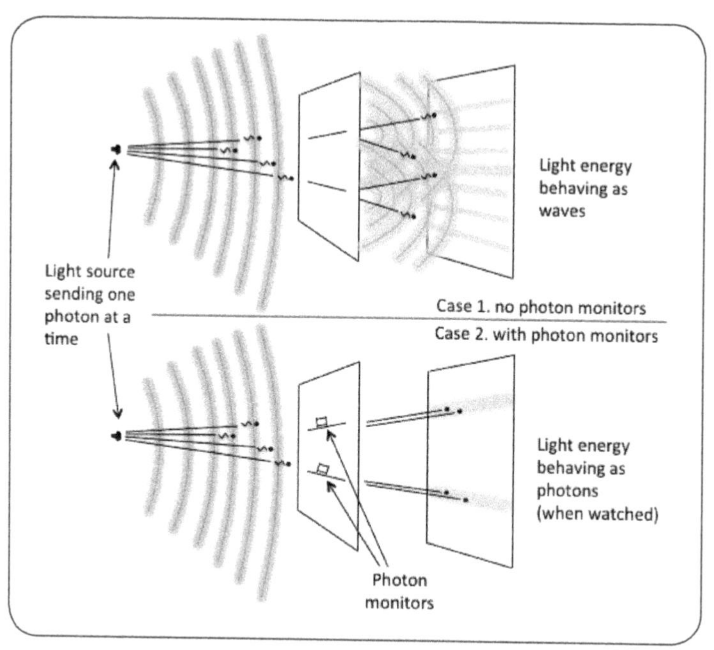

Figure 10-10. The double slit experiment. *When photons or other subatomic particles are sent through a board with two slits, one at a time, the photon or particle behaves like a wave apparently going through both slits (Case 1 - upper portion). But when scientists look to see which slit the particle goes through, it then behaves like a particle just going through one of the two slits (Case 2 - lower portion).*

The conclusions being drawn from the double slit experiment are indicative of many strange new discoveries about how the Universe works at the subatomic level. These discoveries are concluding that subatomic particles don't have actual locations, but rather just the probabilities of locations defined by their associated waves. The waves, scientists are concluding, really are "probability waves" that define the probability of the particle being at a given location. It is only when scientists look for the particle, like with the photon monitors in the double slit experiment, that you "collapse" the probability wave function and determine the location of the particle. In this way, it is the human act of seeing the particle that determines the location. It is human perception, human consciousness, that is creating the reality, at least at the subatomic level.

Scientists have suggested that entanglement may be explained by probability waves. When the correlated characteristic of one particle is determined through experiment (i.e., when its probability wave function is collapsed), the correlated characteristic of the other particle will also be determined, regardless of how far the particles have moved apart. Einstein argued against the interpretations being proposed and looked for explanations that did not seem to violate Special Relativity by implying information traveling faster than speed "c." He was not successful in his arguments, and near the end of his life he wrote a letter to his colleague, David Bohm, stating that after many years of trying, he did not have the slightest idea what kind of elementary concepts could be used to explain the predicted phenomenon.

There were no experiments that were possible during Einstein's life that could verify the correctness of entanglement. However, since Einstein's death, numerous experiments have been conducted and have demonstrated that entanglement does indeed occur. Scientists still do not fully understand how the phenomenon works. Are there things we don't understand about location? Or is it the definition of reality that is unclear?[2]

Wormholes. Wormholes, predicted as possible by Einstein's General Relativity equations, are highly theoretical structures in space-time that are hypothesized to connect multiple black holes. The extreme bending of space-time created by black holes, perhaps complicated by extreme rotations that may characterize newly formed black holes, could deform space-time to such an extent that a passage, called a wormhole, may temporarily be created connecting the black holes.

If wormholes exist, they could connect black holes separated by large distances, thus creating a connection between distant locations in space-time. For example, locations in space-time separated by many light-years could be bridged by a wormhole that is only a few miles, or even a few feet, long. According to physicist Kip Thorne, these would be very short-lived. But if they could be stabilized, they could theoretically allow travel between the remote areas of space-time connected by the wormhole. In this way, it would theoretically be possible to traverse huge expanses of space-time very quickly; much faster than possible through normal space-time.[3] However, since wormholes are created by severely bending space-time to connect the distant black holes, this is more like taking a shortcut, than actually exceeding speed "c."

Some researchers have recently suggested that wormholes may routinely exist at the subatomic level and may even provide an explanation for

2 Like for the double slit experiment, it is difficult to visualize and understand entanglement from a textual description. There are several YouTube videos that do a nice job of describing it visually: www.youtube.com/watch?v=tafGL02EUOA and www.youtube.com/watch?v=ZuvK-od647c&t=304s.

3 See Kip Thorne's book, *Black Holes & Time Warps: Einstein's Outrageous Legacy*. For a fascinating discussion of the theoretical possibilities of wormholes.

the workings of entanglement. While fascinating things to think about, wormholes are completely theoretical. They are based on mathematical solutions of Einstein's equations, and have not been verified to exist.

10.5 The appearance of compressed space or actual compressed space? Things do become shortened under conditions of relative motion and acceleration due to space compression described earlier, but only optically! The real physical size of things is not changed. However, the optical reality of space compression is a real aspect of our Universe.

So what is it? Do things actually get shorter when in high-speed relative motion or do they just appear to be shorter? Are the atoms and molecules of a rocket ship passing by at near the speed of light really getting closer together? The overly simplistic answer is that just the appearance of things become compressed. This is because the compression of space is based on the movement of light; and light is the basis for seeing and for how things appear. When we look at an object moving at near the speed of light, it will look shorter (compressed) because light travels at speed "c" regardless of relative motion between frames-of-reference. This translates into the appearance of compressed space and slowed time, as described earlier.

But a more complete, and more correct, understanding has to do with the fact that space is just one component of space-time. It is only when viewed separately that time appears to slow and space appears to become compressed. If we were able to experience space-time as an integrated whole, we would understand that apparent distortions in time and size are the result of changing perspectives. That is, when in high-speed relative motion, objects appear in a more "edge-on" perspective, making the time component of space-time appear slower and the space component to appear compressed. This is what is really happening.

This is analogous to seeing a book straight-on versus edge-on in three-dimensional space, as shown in Figure 10-11. The size of the book viewed edge-on appears smaller. Artists call this foreshortening. But we aren't confused by it because we are used to changes of perspective viewed in three-dimensional space. We know that the book hasn't really gotten smaller, just that we are seeing it from a different angle. However, when we learn that high-speed relative motion creates similar perspective distortions in four-dimensional space-time, creating the appearance of slowed time and compressed space, we have difficulty appreciating how that works. This is because our everyday relative movements do not approach the speed-of-light. We do not have experience that teaches us how perspectives change for objects in near speed "c" relative motion. The very slow speeds that we experience on Earth are not fast enough to make these distortions perceptible. If we regularly experienced speeds approaching the speed of light, we would recognize that the apparent time slowing and space shortening are caused by changes in perspective. It would be obvious.

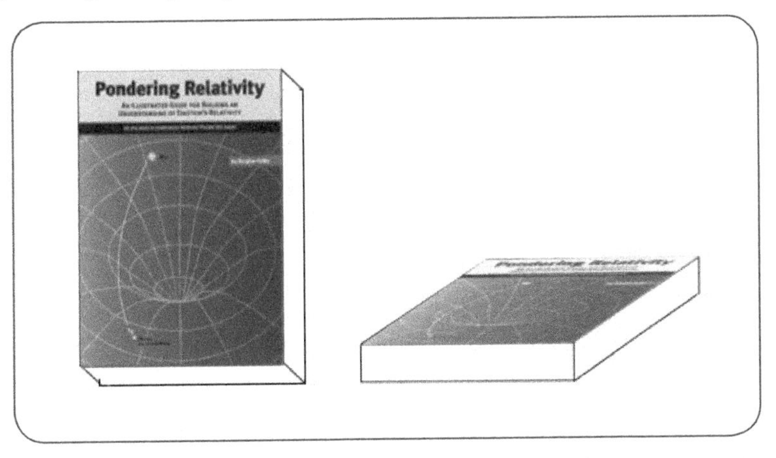

Figure 10-11. Time slowing and space compression result from changes in viewing perspective. *The apparent slowing of time and compression of space at relative speeds approaching speed "c" results from changes in the perspective from which we view space-time. This is analogous to how objects viewed in three-dimensional space project a smaller apparent size when the viewing angle (or perspective) changes.*

One might ask why time slowing and space compression are only significant for relative motions approaching the speed of light? The reason for this is that space and time are integrated in the ratio of 186,282 miles (across the three spatial dimensions) per second (the time dimension). It is only at speeds approaching this level that the changes in perspective become significant. The relative motions that humans routinely experience result in negligible perspective changes that are far too small to make space compression or time slowing noticeable. Our relative motions do not come close to the 186,282 miles per second, the ratio between time and space. We therefore do not experience enough time slowing and space compression to give us a basis in our reality for appreciating how it works.

And isn't there an actual length of objects when they are not moving; when they are not being seen from another frame-of-reference; when they are just in the frame-of-reference where they actually exist? And isn't it just that they appear to be shorter when seen from another frame-of-reference; a reference frame that changes the space-time perspective? Yes, this is exactly the case. An object is longest in the frame-of-reference in which it actually exists, where it is not viewed in motion relative to an observer. This is called the object's "proper length." It is the object's longest length and its actual length. It only appears to be shortened when viewed in relative motion due to the apparent compression of space — resulting from the altered space-time perspective. It's just like the foreshortened image of the 3-dimensional book when viewed in 2-dimensions.

One might conclude that space compression and time slowing associated with moving objects, and the effects of acceleration and gravity on how things appear, are less "real" than measurements made within our own frame-of-reference. This conclusion would be wrong because we exist in a Universe where everything is moving and subject to a myriad of accelerations and gravity. To make sense of our Universe, we have to account for these effects. We have to use the laws of relativity to accurately measure and understand all aspects of the Universe, a Universe that presents itself visually with photons all moving at speed "c" (relative to all frames-of-reference). It's as real as the Universe itself.

And finally, when we talk about time slowing and space compression, either because of high-speed relative motion or gravity (i.e., space-time distortions), we are referring to time and space deviations considered independently from one another. Two people can disagree about distances in space, how fast clocks are going, or even the rate of aging based on their own relative movements. However, all people would agree on the separations in space-time, if we were able to perceive and measure space-time as an integrated thing. It is only because we look at space and time separately that we find disagreements. Of course, it is natural for us to think of, and perceive, time and space as separate because of the fundamental differences in how space and time work — our movements through time being outside of our control and always moving from present to future, while our movements through the spatial dimensions seem more within our control.

10.6 How objects bend space-time. So how does space-time know that there is a gravitational object nearby and that it needs to bend? This section answers this question.

As noted earlier, space-time — when located near large, massive bodies — becomes distorted in a way that makes space seem to be compressed and time to move slowly. This happens because the mass of a large body like the Earth, for example, creates a gravitational field around itself. It is the gravitational field that makes space-time bend.

This is similar to how magnetic fields around a magnet tell the surrounding area where the magnetic field is. Figure 10-12 represents how a bar magnet creates a magnetic field that causes metal filings to line up according to the magnetic field "lines of force." You may remember this from science class. The Earth also creates a magnetic field (in addition to a gravitational field). The Earth's magnetic field is simplistically represented in Figure 10-12 alongside the comparable magnetic field of a bar magnet. It's the Earth's magnetic field that makes your compass needle point north. The compass needle points north because that direction lines up with the Earth's magnetic field, just like the metal filings in science class line up along the bar magnet's magnetic field.

The way magnetic fields are created (for both bar magnets and the Earth alike) is with electromagnetic waves. These are directly related to light waves. In fact, as we have seen, light waves are a form of electromagnetic waves. Electromagnetic waves carry both electrical and magnetic field energy. They have waves and energy packets called photons that travel at the speed of light (in a vacuum), as discussed earlier. These are well understood. Scientists have been able to isolate photons and convert between electrical, magnetic and the combined electromagnetic energy.[4]

4 For a more complete description of electromagnetic fields and waves see the NASA website (missionscience.nasa.gov/ems/02_anatomy.html). This website also has a short description of the relationship between electricity, magnetism and electromagnetic radiation (including light).

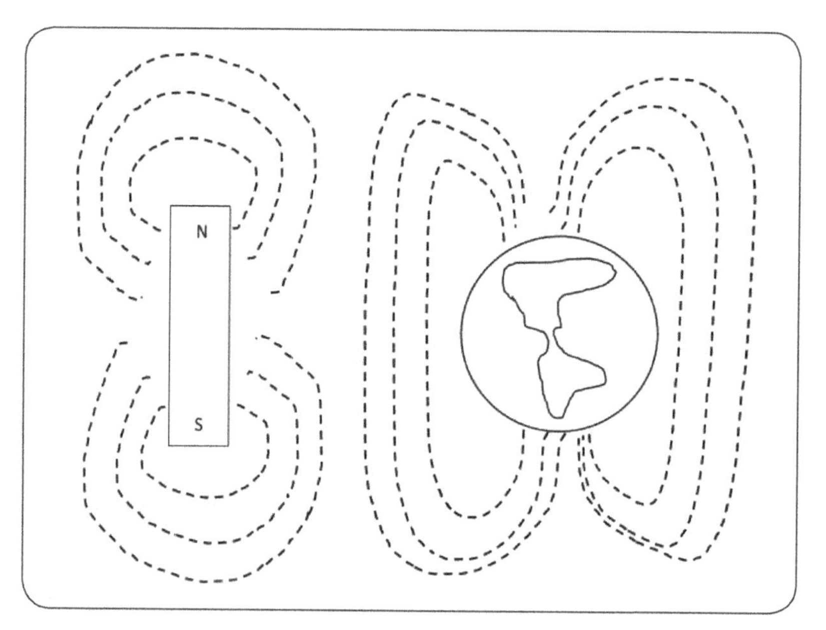

Figure 10-12. How magnetic fields are created. *Magnetic fields work by sending photons, just like for light. The photons travel at the speed of light and inform space-time where magnetic lines of force are.*

Gravity also works by creating "fields" in a manner that is similar to the way magnetic fields are created. Both magnetic and gravity fields are carried by waves with associated energy particles. The energy particles that carry magnetic fields, as we have seen, are photons. The energy particles that carry the gravitational field have yet to be discovered, and therefore are still hypothetical. The hypothetical particles that carry gravity are called gravitons, and according to Einstein's theories are believed to travel at speed "c" like photons.[5] While gravitons are yet to be isolated and confirmed, scientists have recently detected and measured gravity waves. This very recent and exciting discovery confirmed the existence of gravity waves as predicted by Einstein. This is described in Chapter 11 and is another confirmation of the correctness of Einstein's General Theory of Relativity. The confirmation and measurement of gravity waves was first made in February of 2016.

Figure 10-13 presents a simple representation of compressed space-time around the Earth corresponding to the gravitational field created by gravity waves. The gravity waves tell space-time how to bend. As we have seen, it is the geometry of space-time that seems to pull things like apples, spaceships,

5 If gravitons move at speed "c" as believed, this means that if the Sun were to disappear, it would take a short time for space-time to unbend. Specifically, it would take about 8 minutes for space-time near the Earth to unbend, and for the Earth to stop orbiting the Sun.

and people toward the Earth. Like many figures in this book, Figure 10-13 is a fanciful, over-simplification that helps us visualize the effect of more compressed space near Earth, while not accurately showing the underlying mechanisms from a physics perspective.

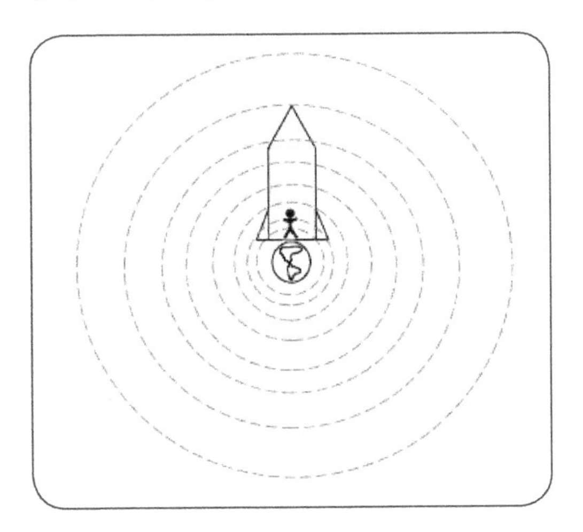

Figure 10-13. How gravity waves work. *Gravity waves create a gravitational field around large massive bodies. It is the gravitational field that bends space-time and makes objects move toward the Earth. The gravitational field created by gravity waves can be visualized as concentric circles with graduated diameters indicating more compressed space-time near Earth, as shown here. The gravitational waves themselves move though space-time in a manner similar to light waves, but with gravitons instead of photons and an actual bending of space-time.*

10.7 The structure of space-time. Historically, space has been described as being empty. It is thus frequently referred to as "empty space." Now that Einstein's theories have combined space with time to form space-time, does that mean it is really space-time that is empty? If not, what is space-time full of? What is space-time made of? What is space-time, really? This section describes.

In this book, space-time is described as the integration of space with time. Time by itself, and space by itself, have no independent existence in Einstein's Universe. It is space-time that provides the underlying framework for the Universe; the thing or fabric that light and other forms of electromagnetic radiation go through. We no longer see space as being made of a rigid three-dimensional aether, but now understand that the Universe is defined by four-dimensional space-time.

So what is space-time made of? What is it that light goes through? What is it that gets bent by gravity? The answer is that scientists really don't know. But there are some discoveries that are providing hints.

First, we have learned that space-time is not empty. It is full of stuff that we can't see using electromagnetic sensors like telescopes. It is full of matter and energy that is invisible; that is, stuff that doesn't interact with photons, so we can't see it. The stars, planets, nebula, and scattered matter that we see throughout the Universe are only a small percentage of what's there. In fact, 70% of the Universe is what's called "dark energy," 25% is dark matter and 4% is regular atoms that are invisible to our sensors. This means that 99% of the Universe is stuff we can't see! The stuff we can see (galaxies, stars, planets, nebula) makes up only 1% of the Universe[6].

The exact nature of dark energy and dark matter is not known. Scientists are able to discern the presence of dark matter and dark energy by the effect they have on the things we can see throughout the Universe. Scientists know, or at least presume, that dark matter exists because of its apparent gravitational effects on the movement of the stars and galaxies that we can see. That is, there is more gravitational bending of space-time than can be explained by what we see in the visible Universe. Something must be out there to create the extra, unexplained gravity. Scientists call this something "dark matter."

Similarly, we know that dark energy is there because of its apparent effect on the movements of the galaxies and galaxy clusters that we see. Dark energy is even more amazing and surprising than dark matter. Instead of pulling things together like gravity, dark energy seems to be pushing things apart!

Not only is the Universe expanding, perhaps resulting from the Big Bang, the explosion-like beginning of the Universe that started it all some 13.8 billion years ago, but the Universe appears to be expanding at an accelerating rate! This is most obvious at the furthest reaches of the Universe where galaxy clusters appear to be moving apart at an ever-increasing (accelerating) speed. While the galaxies within the galaxy clusters themselves seem to be held together through normal gravitational attraction (the bending of space-time), the larger clusters of galaxies seem to be accelerating apart. Something is pushing the galaxy clusters apart, overcoming

6 See Abrams, Nancy Ellen and Primack, Joel T. (2011). *The New Universe and the Human Future: How a Shared Cosmology Could Transform the World* for a discussion of what the Universe is made of.

the gravitational effects that should be present. This energy source is not understood, but is called dark energy.

An important question raised by an expanding Universe, especially one whose expansion is accelerating, is whether more space-time is being created or if the existing space-time is being stretched. And this begs the question, what is space-time? What is space-time made of? What is it that the expanding Universe seems to be making more of or stretching? Again, scientists don't really know. But they know it is something real because waves of gravity, as well as electromagnetic energy waves (like light), propagate through it. Scientists have been measuring electromagnetic waves (e.g., light waves) and the associated photons for over a century. These waves propagate through space-time. However, because photons have no rest mass (only momentum) their effect on space-time is minimal and they pass through it without changing it. They are affected by it as we have seen (e.g., the bending of light near gravitational sources), but otherwise just seem to go through it. Gravity waves, on the other hand, change the shape of space-time as Einstein predicted and as scientists have recently confirmed.

It was only very recently (February 2016) that scientists have been able to detect and measure gravity waves moving past and through the Earth. Gravity waves are a more direct indicator of the nature of space-time because they have a measurable effect of bending it as they pass through. This recent discovery will be described in Chapter 11. The gravity waves were detected by measuring the associated bending of space-time reflected in changes to light waves, also passing through space-time. While the question of what space-time is made of is still up for debate, the fact that scientists can now measure how it deforms as gravity waves pass through it helps scientists to study its nature. So stay tuned.

10.8 $E=mc^2$. Most people associate the formula $E=mc^2$ with Einstein and relativity. This formula defines how energy and mass are interrelated and exchangeable. It mathematically defines the equivalence between mass and energy, which is an important principle associated with General Relativity. This is explained in this section.

Einstein's famous equation had very little to do with the development of his first theory, the Theory of Special Relativity, but grew out of it. The equation was not in Einstein's papers in which he first presented his ideas for Special Relativity, but emerged when questions that were raised by the theory were addressed. Einstein considered it to be one of the most important consequences of the Special Theory of Relativity.

The formula $E=mc^2$ was derived when the Special Theory of Relativity was applied to matters of calculating the total energy (E) associated with the combined inertial masses of multiple objects (m) when measured across different frames-of-reference. Once space and time were recognized as an integrated thing (space-time), the combined inertial energy of objects moving relative to each other became associated with their combined masses and their relative velocities through space-time (not just through space). Therefore, the constant that combines space with time (speed "c") became a central variable for measuring the combined inertial energy. The famous equation emerged.

The formula seems simple. However, the derivation is complicated because, when objects are in motion relative to each other, the masses are relativistic masses. Relativistic masses of moving objects combine the objects' rest masses with the mass equivalent from the kinetic energy of motion. Mathematically, when relative speeds approach speed "c," the kinetic energies approach infinity. And this means that the relativistic masses also approach infinity, which can't be achieved. This is how the limit of speed "c" is preserved.

These more complicated formulas are beyond and outside the scope of this book. It is enough to say here that Einstein's insights about relative motion, the integration of space with time, and equivalence of mass and energy led to the understanding that objects moving with relative speeds approaching speed "c" are associated with infinite amounts of kinetic energy that can't be achieved. See Robert Resnick's "Introduction to Special Relativity" for an illustrated mathematical description.[7] Beyond the math and physics, intuitively this makes some sense. It would require an infinite amount of energy to move objects faster than speed "c" because that would violate the

7 See *Introduction to Special Relativity* by Robert Resnick (pages 119 – 131, especially pages 122 – 123). Also Einstein's introductory book on relativity, *Relativity: The Special and General Theory* (page 44), provides a discussion of the ramifications of Special Relativity for combining inertial energies.

very structure of space-time that combines space with time in the ratio of speed "c."

Building on the foundation of Einstein's theories of relativity, the simple equation $E=mc^2$ shows that there is a direct and intimate relationship between mass and energy and that, in fact, the two are interchangeable. This in itself does a lot to help us make sense of how the Universe works. It says that mass and energy are two aspects of one thing — a thing that could perhaps be more accurately called mass-energy. The formula says that energy (E) is equal to the mass (m) times the speed of light (c) squared. Most importantly, the formula says that mass can be exchanged for energy and energy can be exchanged for mass (i.e., matter), according to this relationship. Mass and energy are two manifestations of a single underlying reality, and each can be exchanged for the other in accordance with the relationship $E=mc^2$. Not only did Einstein combine space and time into a single thing, space-time, he combined mass and energy into an integrated thing that can be thought of as mass-energy.

The most direct implications of $E=mc^2$ are for understanding the underlying mechanism involved with producing energy. This applies to all energy production operations like nuclear power plants or coal-fired power plants, but also includes all actions that produce energy. This includes such things as burning a candle, burning wood in a campfire, or even taking energy from a compressed spring like in an old-fashioned windup wristwatch. In every case, a tiny amount of the mass is converted into energy. If we could very accurately measure the mass of all the material involved in generating the energy — for example, the wood of a burning fire, the oxygen used during the burning, and the carbon dioxide given off, etc. — we would find that a very tiny amount of mass is lost. We would also find the amount of mass lost compared to the amount of energy produced is determined by the formula $E=mc^2$. While the amount of mass lost in a burning fire is way too small to measure, the relationship between mass and energy has been measured and verified with nuclear reactions in which small amounts of enriched uranium are used as fuel and very large amounts of energy are produced.

Most importantly, though, with respect to the focus of this book, $E=mc^2$ has important implications for how the Universe itself is made. Before Einstein, the belief among physicists, and people in general, was that the Universe was made of space, mass and energy. The belief was that these three components were completely separate things, and all three were completely separate from time. As we have seen, Einstein successfully combined space and time into a single entity, called space-time; and with his equation $E=mc^2$, he combined mass (matter) with energy and showed how these two are completely interchangeable. He changed how we understand the Universe in fundamental ways.

10.9 Limits of relativity's scope and competence. Relativity greatly enhanced our understanding of the Universe, but there are limits to what relativity can tell us. This section explores these limits.

Science and our understanding of the Universe is continually evolving and growing. Einstein's theories of relativity have played a major role in expanding this understanding. Today, it forms a central, foundational place in modern physics. It brought together several areas of physics that were previously either poorly understood or only understood as isolated physical laws that lacked deep knowledge of underlying principles and causes. It established a scientific basis for explaining the behavior of a wide range of physical phenomena observed. Today it is the go-to theory for bringing together our understanding of diverse phenomena ranging from electricity, magnetism and gravity, to astronomy and cosmology.

While relativity provides a scientific understanding that integrates across this wide range of areas within physics, there are some areas of physics, and aspects of how the Universe works, that relativity has been unable to explain. Basically, relativity works well for describing the behavior of large things, like the motion of galaxies, stars, planets, rocket ships, and even bowling balls and sticks and stones; and it helps us understand what makes electricity and its partner magnetism work. However, relativity has been less successful in describing how the Universe works at the very smallest scales, such as deep inside atoms and molecules, and inside black holes where extremely large energies and gravitational forces are present within a very tiny space. For these extremely tiny and high-energy realms, other models, theories and laws have worked better, although with varying levels of success.

One successful theory for describing physics at the smallest scales is called Quantum Theory. Quantum Theory describes the nature and behavior of the

Universe at atomic and subatomic levels. It identifies quantum particles as the smallest constituents of matter and energy. These particles have defined energy levels and behave in a wave-like manner where the wave frequency specifies the amount of energy present. These quantum particles are really more like discrete energy packets that behave like waves as described earlier for light waves. Each energy packet represents the smallest amount of energy possible for the type of matter or energy being defined. Each discrete energy packet is called a quantum of energy, thus the name Quantum Theory. Photons, described earlier, are an example of energy packets defined by Quantum Theory.

Quantum Theory has proven to be very good at describing the natural behavior of things at the subatomic level. Many quantum particles have been discovered and studied, while others are still only hypothesized. Attempts to fully define Quantum Theory and how it works have led to a proposed categorization of the full range of quantum particles that would be associated with all known matter and energy types. This categorization is called the "Standard Model." Much work remains to be done to fully define and understand reality at the subatomic level.

Research to understand how quantum particles work has led to some novel, even strange, conclusions about the Universe and reality itself. For example, scientists have found that the position and physical properties of quantum particles are defined by mathematically-based "probability waves" rather that classically defined physical movements of solid particles. Some scientists during Einstein's life proposed conclusions about the very nature of existence, in which reality is created by the act of observing. That is, the act of human observation plays a central role in defining what is real, and what exists. Even though Einstein played a major role in developing Quantum Theory, he never fully accepted all of these ramifications and conclusions about the nature of reality that were being drawn. He famously said, "God doesn't play dice with the Universe." Many of these conclusions are now widely accepted, while others are still controversial.

The most prominent hypotheses for explaining quantum-level reality and behavior define vibrating strings of energy that vibrate in up to 11 separate dimensions.[8] These proposed hypotheses are referred to as "String Theory"

8 See *The Elegant Universe: Superstrings, Hidden Dimensions, and the Quest for the Ultimate Theory* by Brian Greene for a discussion of string theory.

and are still hypothetical and controversial. According to the String Theory, vibrating energy strings combine to make up and define the behavior of the many subatomic particles. Much work remains to be done and some scientists have questioned the correctness of the string theory hypotheses.[9]

While relativity does not provide complete descriptions of how things work at subatomic scales, Einstein contributed to efforts that tried to expand relativity to work at the subatomic levels. These efforts tried to combine General Relativity with the principles of Quantum Theory. He was not successful in this effort during his lifetime, but work continues in this quest, some building on the ideas he proffered. For example, hypotheses that combine relativity with Quantum Theory, collectively called quantum gravity, continue to be explored.

At this time there is no single accepted theory that can describe both large and small-scale physics. But work continues, and new and better tools for pursuing this area of research (such as the new Hadron collider near Geneva, Switzerland) are now available. The potential for success in understanding the subatomic world is increasing.

Despite exciting successes and developments in Quantum Theory, quantum gravity, and string theory, these areas are all beyond the scope of this book. This book focuses solely on relativity and its success in describing our Universe on large scales. References for further reading in these other areas are provided in the Bibliography.

---

9 See *The Trouble With Physics: The Rise of String Theory, the Fall of Science and What Comes Next* by Lee Smolin.

Chapter 11
Where's the Proof?

This chapter addresses the following topics and questions:

11.1 The questioning scientific community. *How did the scientific community react to Einstein's theories of relativity?*

11.2 Verification of Einstein's predictions. *What verifications did Einstein propose for demonstrating the correctness of his theories? Were they able to be tested? What were the results?*

11.3 Modern experiments that have measured the effects of relativity. *What recent experiments made possible by modern technology have been conducted to verify the correctness of relativity?*

11.4 Putting relativity into practice. *Has our understanding of relativity made modern technology possible? Are there breakthrough technologies enabled by relativity?*

When Einstein first proposed his relativity hypotheses, they represented a very new way of looking at the Universe, and met with initial skepticism and even some resistance. At the time, it was difficult to verify his ideas experimentally and so they were not immediately accepted. Eventually, experiments were conducted and scientific evidence was developed that supported his ideas. They now form a foundation for how we understand the Universe. This chapter describes how the scientific community responded to Einstein's ideas and summarizes some of the scientific evidence that eventually led to widespread acceptance of both Special and General Relativity.

11.1 The questioning scientific community. The immediate reaction to Einstein's theories of relativity were mixed. At first, there was some resistance and skepticism, but eventually there was widespread acceptance.

Einstein's theories when first proposed ran counter to the status quo of theoretical physics of the time; a status quo based on the "classical physics" of Isaac Newton. This was a physics that assumed that space had a physical structure defined by an aether through which objects moved and light waves propagated. It maintained that all motion could be measured in relation to the aether; and that the measurement of motion was simply a matter of determining the location of objects in space across different points in time. Time was believed to be a separate entity; a dimension unto itself that moved from past to future, but had nothing to do with space. And it was a physics that believed that planets moved in their orbits because they were pulled by a force called gravity; a force generated by large bodies like the Sun, but one whose mechanism was completely unknown. Newton had figured out how much gravitational force was generated by the Sun based on the distance and orbital paths of the planets, but had no idea how the force worked and how it was communicated to the planets.

Einstein's relativity hypotheses completely overturned these accepted scientific beliefs. His idea that clocks in relative motion will run more slowly and that "measuring sticks" (rulers) in motion will be shorter than those at rest seemed strange, indeed, to many. And Einstein's contention that time is really a fourth dimension forming four-dimensional space-time, and that gravity works simply because space-time is bent by large massive objects, took some time to sink-in and to gain acceptance. Even the notion that objects cannot exceed the speed of light relative to other objects, a notion for which there was scientific evidence, was not immediately accepted. Add to this the fact that Einstein was employed as a junior patent clerk when he first published his ideas on Special Relativity rather than having an appointment as a professor at a university. This led to even more skepticism about his strange ideas. The lack of initial acceptance of Einstein's theories by the scientific community is highlighted in Figure 11-1.

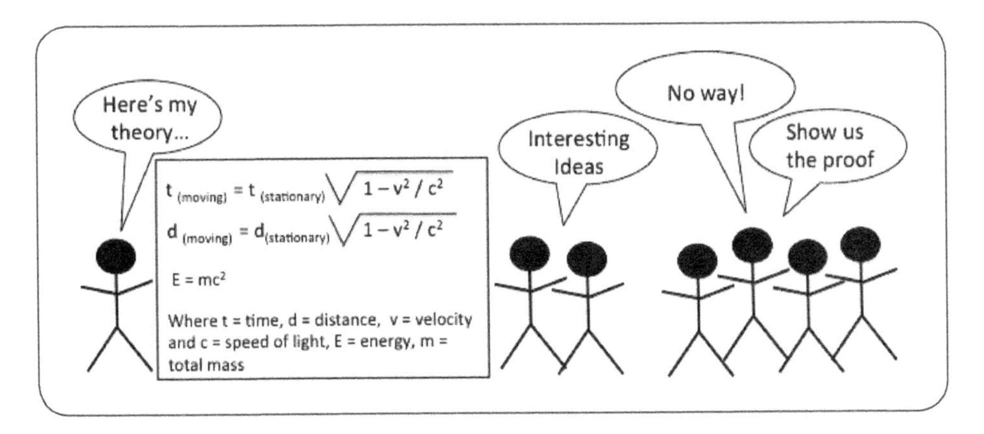

Figure 11-1. At first, most scientists didn't accept Einstein's theories.

In spite of his lowly position as a patent clerk, Einstein published his initial proposals about Special Relativity in the respected German journal "Annalen der Physik" in 1905. These ideas elicited some discussion in Germany, but were all but ignored elsewhere. Most physicists who paid any attention at all reacted strongly against the new theories as being impractical and inconsistent with existing scientific knowledge. But Einstein persisted and by 1908, at least some scientists began to accept the principles he promoted. In 1908, mathematician Hermann Minkowski reacted positively to Einstein's idea that space and time are actually a single integrated reality. Minkowski famously remarked, *"henceforth space on its own and time on its own will decline into mere shadows, and only a kind of union between the two will preserve its independence."* In fact, it was Minkowski who initially coined the term *"space-time."*

By 1909, Einstein was able to leave the patent office and secure a position as a professor of theoretical physics at the University of Zurich. In 1910, he secured a full professorship in Prague in Czechoslovakia. By 1911, the implications for expanding Special Relativity to cover acceleration and gravitation were beginning to be discussed. It was in 1911 that Einstein first published calculations based on his Equivalence Principle that predicted effects on orbiting bodies that were eventually explained by General Relativity, the broader theory of relativity that accounted for conditions of acceleration and gravity.

But even after the General Theory of Relativity was published and began receiving widespread support among physicists and astronomers, there were some respected scientists who still argued in favor of an aether and proposed various approaches to account for the shortening of objects that are in relative motion and the bending of light around large bodies like the Sun. For example, they reasoned that objects would shorten when pushing against the aether as they moved through it and the aether density would vary near large bodies like the Sun. And they reasoned that the constant speed of light could result when large bodies like Earth dragged a portion of the aether along as they moved through.

Eventually, a large enough body of evidence was developed that led to almost universal acceptance of relativity, and of Einstein, who eventually achieved nothing less than celebrity status throughout scientific circles and among the larger general populations. The ultimate acceptance of Einstein's theories is highlighted in Figure 11-2.

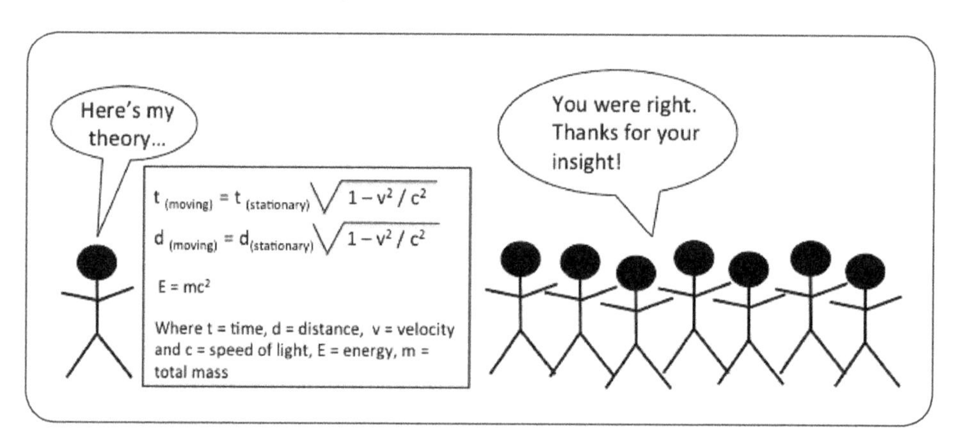

Figure 11-2. Eventually, Einstein was congratulated for his work.

It is not unusual for new ideas to be questioned. This is how science works — when one scientist proposes a new theory or idea, it has to be demonstrated and verified using scientific measures and experiments before it is accepted. This is especially true when very novel ideas are proposed like those of Einstein.

The body of evidence that ultimately supported relativity is summarized in the following sections.

11.2 Verifications of Einstein's predictions. General Relativity made three predictions that Einstein said would demonstrate the correctness of relativity. This section describes the verification of these predictions.

Eventually experiments were done and observations made that showed that Einstein's theories accurately describe how the Universe works. The first of these were predictions made by Einstein himself. Specifically, his theories postulated: (1) that the previously unexplained orbital path of Mercury will be shown to conform to the equations of General Relativity; (2) that light from distant stars will be shown to deflect when passing near a massive object like the Sun; and (3) the frequency of light will be reduced (toward the red end of the visible light frequency spectrum) when the light moves away from the compressed space-time near the Sun. The results of experiments attempting to verify these predictions are summarized below.

Unexplained orbital path of Mercury. It had been known for many years that the orbital path of the planet Mercury, the planet nearest to the Sun, deviated slightly from the path that would be predicted by Newton's equations. Specifically, the longest axis of the orbit shifts slightly with each orbit as shown in Figure 11-3. This means that the longest axis of Mercury's elliptical orbit shifts 43 seconds every 100 years, much more than predicted by Newton. There was no explanation that could account for this deviation until Einstein applied his equations of General Relativity. To Einstein's extreme delight, his equations for relativity exactly matched the measured orbital path of Mercury.[1] This is because Mercury's orbit follows the contours of space-time deformed by the Sun as described by General Relativity equations, and is not the result of a simple gravitational force as defined by

1 These results are summarized in Appendix C of Einstein's book titled *Relativity: The Special and General Theory.* Also see Jeffrey Crelinsten's book, *Einstein's Jury: The Race to Test Relativity,* for a summary of the growing body of evidence that overcame the initial skepticism surrounding Einstein's theories.

Newton's equations. The eccentricity and shifting of Mercury's orbit shown in Figure 11-3 are exaggerated to better illustrate the effect.

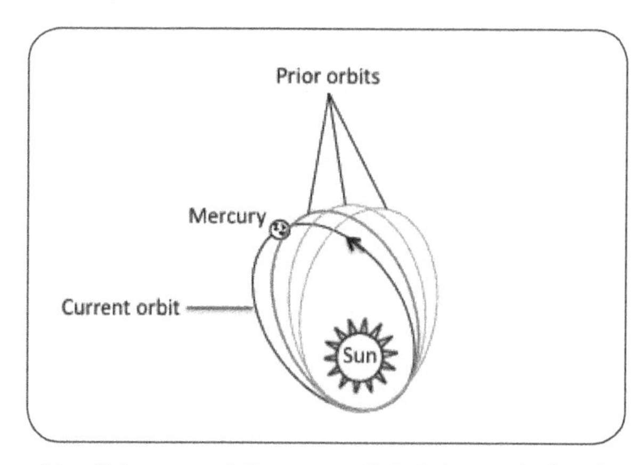

Figure 11-3. The orbit of Mercury exhibits a small shift due to the bending of space-time near the Sun. *The axis of Mercury's orbit shifts slightly due to the effect of General Relativity. This effect is exaggerated here to illustrate the concept. Before Einstein's General Theory of Relativity this small shift could not be explained.*

Gravitational bending of light. Einstein's General Theory of Relativity, published in 1915, predicted that light passing a massive object like the Sun would deflect slightly because of the compression or bending of space-time in that area. This was something that was possible to verify through experiment at the time. But it was only possible to do this during an eclipse of the Sun when nearby stars were visible. The next eclipse was to occur in 1919 and an expedition to measure whether Einstein's predicted effects actually occurred was planned. Although complicated by World War I, Sir Arthur Eddington, a prominent astronomer of the time, led the expedition that made the measurements. Two observation sites were selected (Brazil and West Africa) in case one was obscured by weather or was unsuccessful for any other reasons. Measurements at both sites were successful and both confirmed the predictions of Einstein's General Relativity equations.[2] It was this experiment that firmly established Einstein's General Theory of Relativity as accepted science and catapulted Einstein into celebrity status. This effect is summarized in Figure 11-4.

2 These results are summarized in Appendix C of Einstein's book titled *Relativity: The Special and General Theory*. Also see Jeffrey Crelinsten's book, *Einstein's Jury: The Race to Test Relativity*, for a summary of the growing body of evidence that overcame the initial skepticism surrounding Einstein's including a detailed description of the difficulties involved in making the 1919 measurements.

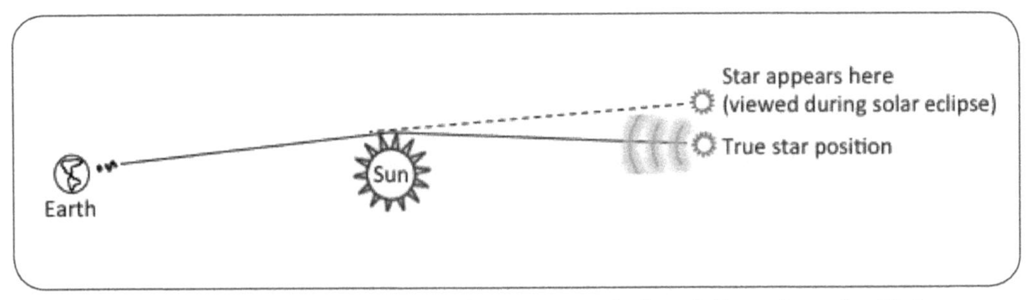

Figure 11-4. Deformed space-time near large massive bodies deflects the path of light. *Einstein's theories predicted that large massive bodies, like the Sun, would bend space-time such that the path of light near the massive body would appear to curve. This figure illustrates this effect that was confirmed by Sir Arthur Eddington's measurements in 1919.*

This effect has been demonstrated repeatedly over the years. It can even result in seeing multiple images of distant objects whose light is bent when it takes multiple paths around very massive galaxies or other massive objects. This effect is illustrated in Figure 11-5. This effect is called "gravitational lensing" because the gravitational bending of space-time acts like a lens and bends the light passing through. This sometimes results in seeing an object's light appear as a ring as it finds many pathways around the intervening galaxy or massive object. Such rings are called Einstein rings in honor of Einstein whose theories predicted the effect. An example of an Einstein ring is shown in Figure 11-5.

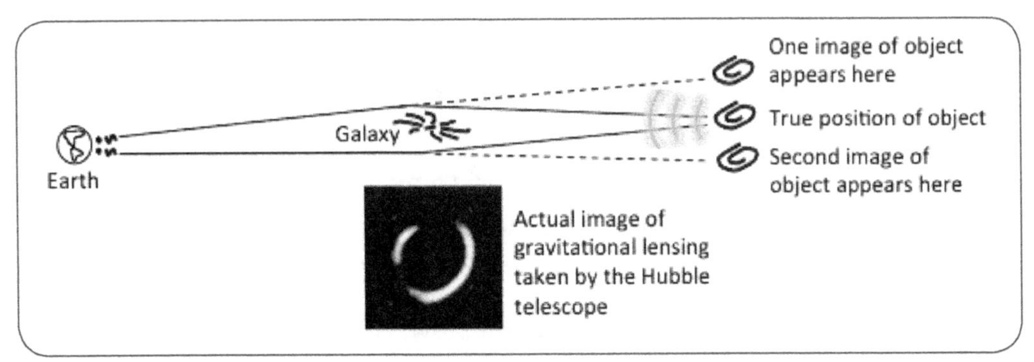

Figure 11-5. Deformed space-time near large, massive bodies can create multiple images of a single object. *Frequently space-time is so extremely deformed around massive objects that multiple images of more distant objects are seen as light finds multiple pathways through the deformed space. This can even create a visual ring around the massive object[3].*

Redshift of light's frequency due to a massive star. Einstein predicted that the frequency of light given off by stars would experience a "red shift" as the light moves away from the star. By knowing the absorption frequency

3 http://hubblesite.org/newscenter/archive/releases/2008/04/image/c/format/web/

patterns associated with chemicals near the Sun, it is possible to measure the amount of "red shift" experienced by light reaching us from the Sun. This is summarized in Figure 11-6. Experiments confirming this prediction have been conducted. These results, as of the time Einstein reported his conclusions, were still uncertain. Some experiments concluded the effect was demonstrated, while others were inconclusive, leaving uncertainty about the effect.[4] The results of later experiments agree with Einstein's prediction and provide more evidence for General Relativity. These experiments are summarized in the next section.

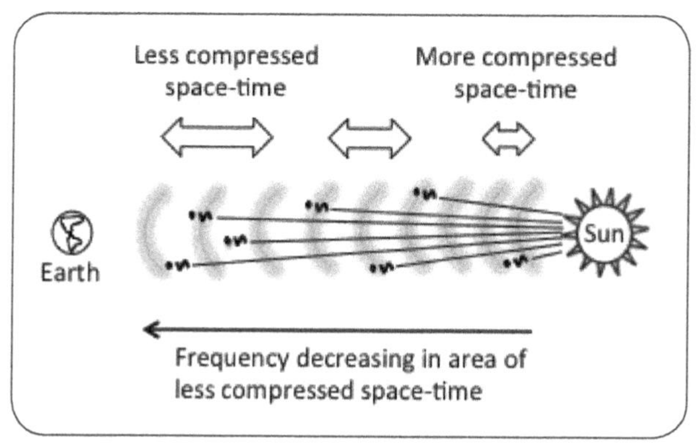

Figure 11-6. The frequency of light coming from the Sun has been shown to decrease as it reaches Earth. *This shift in light's frequency is in agreement with the predictions of Einstein's General Theory of Relativity.*

11.3 Modern experiments that have measured the effects of relativity. Recent experiments that were not possible at the time of Einstein have since been conducted to add additional verification of the theories of relativity. This section describes some of them.

Clocks in relative motion found to run slow. A central premise of Einstein's theories of relativity is that time will move more slowly for observers traveling relative to each other *("moving clocks run slow")*. In 1971, an experiment was conducted to find out. Two very accurate atomic clocks were used to measure this. One of the clocks was loaded on a high-speed jet that circled the Earth at a very fast speed. The other clock remained on the ground. The airplane then landed so the clock that circled the Earth

4 These inconclusive results are summarized in Appendix C of Einstein's book titled *Relativity: The Special and General Theory*. Also, see Jeffrey Crelinsten's book *Einstein's Jury: The Race to Test Relativity* for a summary of the growing body of evidence that overcame the initial skepticism surrounding Einstein's theories.

rejoined the other clock on the ground, similar to the situation experienced by Einstein's hypothetical twins. When the times on the two clocks were compared, it was found that the clock that circled the Earth compared to the clock that stayed on the ground, ticked off less time, thus confirming the predictions of Special and General Relativity.[5] This experiment is illustrated in Figure 11-7.[6] These results have been successfully repeated in numerous subsequent studies.

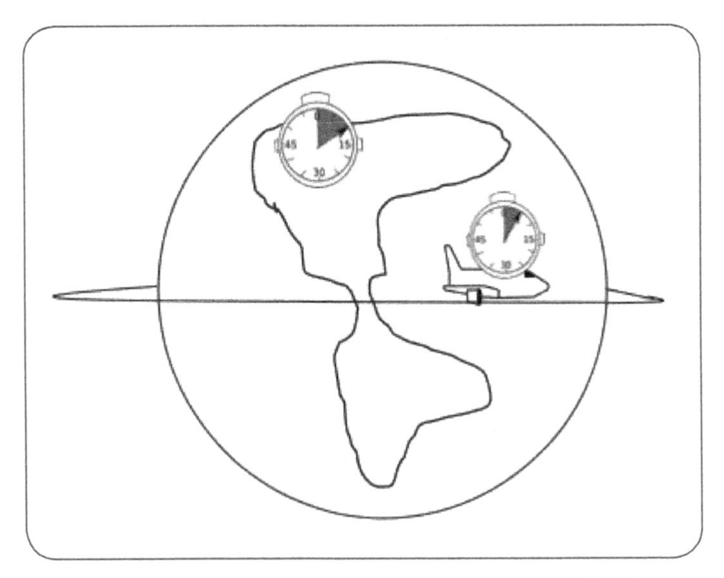

Figure 11-7. Time slowing on moving clocks has been demonstrated. *When an atomic clock was flown in a high-speed aircraft, it was found to run more slowly than an identical clock on the ground. The time slowing was in agreement with the predictions of Einstein's Special and General Theories of Relativity.*

High-energy, radioactive quantum particles decay more slowly when traveling at near the speed of light. Recent research was conducted to further verify that time moves more slowly for things traveling at near the speed of light. Fast moving muons (tiny quantum particles similar to photons) were used in this experiment. muons are created in the upper atmosphere when radiation from the Sun hits other particles in Earth's upper atmosphere. The muons travel toward the Earth's surface at near the speed of light after being created. These radioactive particles decay at a very fast and known rate. They decay so fast that they would not be expected to

5 Note that because the orbiting clock was the one that returned to Earth and joined the other "stationary" clock, it was this orbiting clock that moved relative to the clock on Earth and therefore ran slow. Also, note that the effects of the higher altitude on time passage were minimal for the relatively low gravity of Earth and the low altitude of the planes. It was therefore not a factor in the experiment.

6 The results of this experiment were summarized in two articles by Joseph Hafele and Richard Keating that appeared in *Science* in 1972.

survive their trip all the way down to Earth's surface unless their clocks were running slow due to the effect of Special Relativity. Measures at the surface of the Earth have found that nearly as many muons reach sea level on Earth as are detected at the top of Mount Washington, 6,300 feet above sea level. Only the effects of Special Relativity could explain how so many muons were able to reach Earth's surface before decaying. If the muons were experiencing normal time passage, most would have decayed before reaching Earth. This provides strong additional confirmation that time is slowed and space shortened for the fast-moving muons due to Special Relativity.[7] This experiment is illustrated in Figure 11-8. These effects have since been verified in other moun atmospheric experiments and in particle accelerator studies.

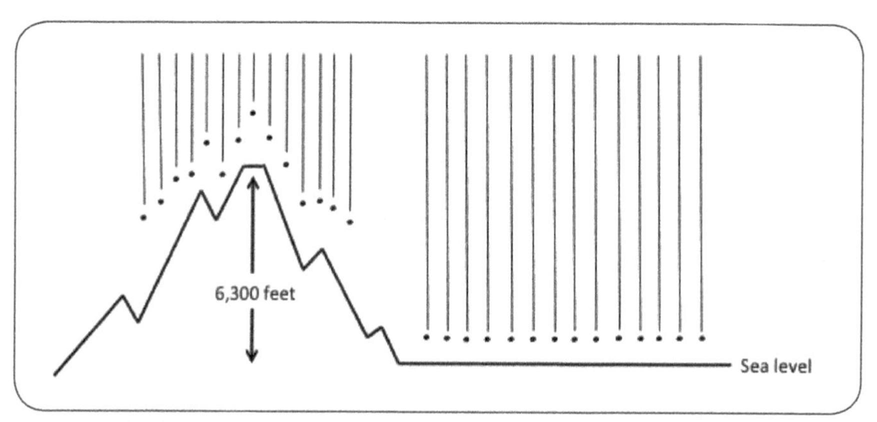

Figure 11-8. The decay rate of high-energy radioactive particles moving at near the speed of light is slower due to Special Relativity. *An experiment compared the number of muons detected at the top of Mount Washington compared to at sea level. Because muons decay very quickly, very few would reach sea level unless their time is slowed by Special Relativity. Since most muons made it to Earth, the experiment demonstrated slowed time experienced by the high-speed muons.*

Electromagnetic frequency shift found due to the compression of space-time near Earth. A central premise of Einstein's theories of relativity is that space-time is more compressed nearer a massive body, like the Sun or Earth. Scientists were able to measure wave frequency changes to radiation sent toward the Earth from the top of a 74-foot tower, and similar changes in the opposite direction for radiation sent from the ground toward the top of the tower. The experiment was conducted in 1960 at Harvard University and has been repeated many times since. The experiment measured selected atomic nuclei that emitted radiation at a known wavelength in the downward direction

7 These results are reported in an article by David Frisch and James Smith published in the *American Journal of Physics* in 1963

and gamma rays in the upward direction. A difference of one thousandth of a trillionth of a wavelength was found verifying Einstein's prediction to within 10%. The experiment was repeated with better equipment in 1964, obtaining results that were within 1% of predictions.[8] This difference reflects compression of space-time nearer Earth, just as Einstein predicted. Figure 11-9 presents a simplified illustration of this experiment.

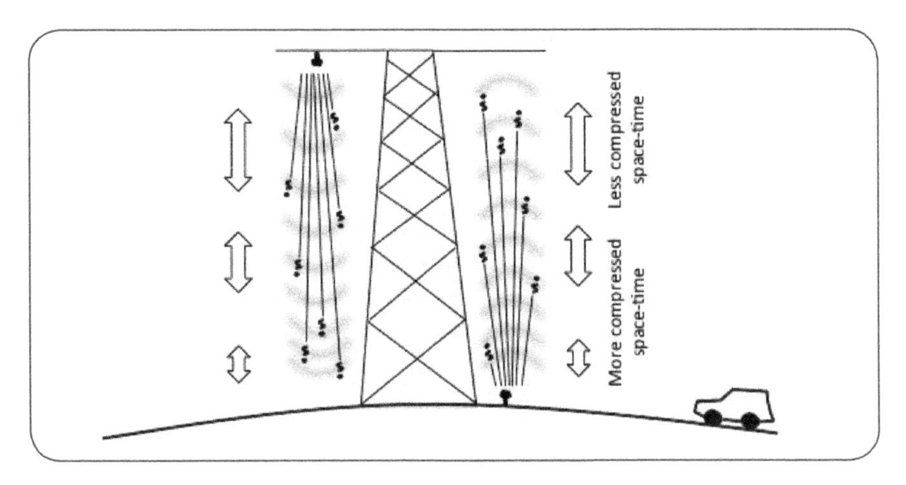

Figure 11-9. Space-time compression closer to Earth's surface was demonstrated. *In this experiment, the wavelength of electromagnetic radiation sent from the top of a tower toward Earth and from the bottom of the tower toward the top of the tower was evaluated. A small but measurable difference in wavelength was detected indicating greater compression of space-time nearest Earth, verifying the predictions of General Relativity.*

Gravity waves detected. Einstein's theories predicted that gravity is carried through space-time with gravity waves. These waves, hypothesized by Einstein, are similar to light waves but work by bending space-time itself as they move through. It's kind of like waves on the ocean bend the surface of the water. But gravity waves bend space-time in all four dimensions as they move away from the mass that created them. The modulations are extremely small and very difficult to detect. Their detection, therefore, requires very sensitive, specialized sensors. Even with sensitive sensors, very large gravitational events are needed to trigger gravitational waves strong enough to be detected.

The sensors that have been designed to detect gravity waves measure changes in space-time when gravity waves pass through. Special large L-shaped sensors, called LIGO (Laser Interferometer Gravitational-Wave Observatory) have been built for this purpose. Two LIGO sensors have been

8 These results are described in an article by R.V. Pound and J.L. Snider published in the *Physical Review Letters* in November of 1964.

installed at two separate locations to make sure that any detection is due to a gravity wave passing through, and not just local vibrations that might happen near the sensors, like a truck driving by for example. The two LIGO sensors were installed in Louisiana and Washington State, respectively. Only when both LIGO sensors make a detection can scientists be sure that it is from a gravity wave. More gravitational wave sensors are planned for deployment in space and elsewhere on earth.

The LIGO sensors work by detecting small variations in the lengths of the two identical sensor arms when a gravity wave passes through, bending space-time. An identical laser beam is sent down each arm and reflected back from the end of the arm by a mirror. When the lengths of the arms change in response to a gravity wave, the distances that the identical laser beams travel also change. Because the wavelength of the identical laser beams in the two arms is the same, the changed relative length of the arms can be sensed when the laser wavelengths become out of phase. This allows LIGO to detect tiny changes in the length of the two arms due to the passing gravity wave. Figure 11-10 illustrates how the LIGO gravity wave sensors work.

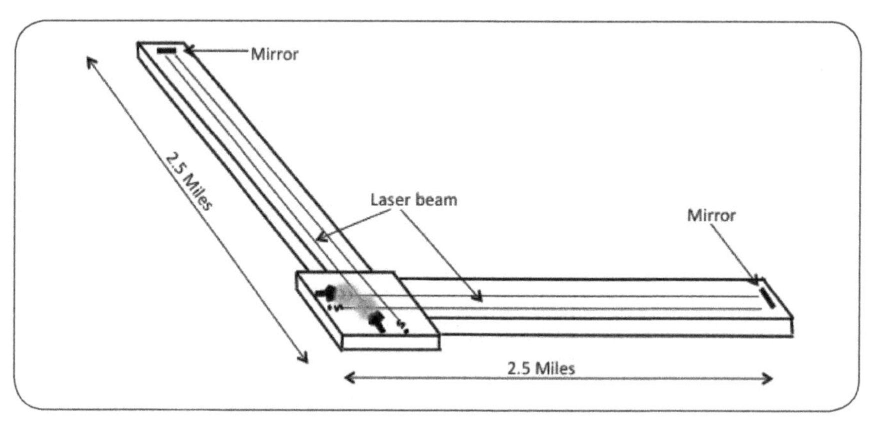

Figure 11-10. LIGO Gravity Wave Detector. *The LIGO gravity detector sends laser beams along two identical sensor arms. When gravity waves come through, they distort space-time and momentarily change the length of the two arms. This causes the laser beams to travel different distances and become out of phase, thus detecting the gravity wave.*

In September 2015, the LIGO sensors in Louisiana and Washington State both simultaneously detected identical small deviation patterns in the lengths of the sensor arms. These results were reported in early 2016.[9] A second detection of a similar event was made in December 2015 and reported in

9 These results were reported by B.P. Abbot and over 50 other authors in the February 2016 issue of *Physical Letters*.

June 2016.[10] Since the measured gravity waves were identical and happened at both LIGO locations, scientists are confident that the deviations were caused be a passing gravity wave, and not just to local vibrations. This simultaneous detection indicated not only the passing of the gravity wave, but actually was able to measure the shape and strength of the gravity wave as it passed through. Amazingly, scientists were able to determine that the wave was created by an extreme gravitational event created when two black holes merged to form a single larger black hole! Scientists have estimated that the event took place one billion light years away. This is the latest scientific confirmation of Einstein's General Theory of Relativity. These results were also widely reported in popular media. An even more sensitive array of gravity wave sensors is being deployed in space between the Earth and Sun to listen for gravity waves. So more information about how gravity waves work is forthcoming.

11.4 Putting relativity in practice. The ultimate test of relativity's correctness is in its application to modern uses. Beyond scientific experiments, modern conveniences are applying relativity to make our lives better. The impacts of relativity on modern life range from nuclear energy that powers many of our cities to modern astronomy that uses the effects of relativity to make sense out of the cosmos. The best example of the application of relativity for improving our everyday lives is GPS, described in this section.

Modern GPS (Global Positioning System) devices have become commonplace today. A key to their success is their accuracy, allowing determination of locations on Earth to within a few feet. If the effects of relativity were not accounted for, the accuracy of GPS would eventually be off by miles.

The effects of relativity on GPS are two-fold. First, GPS satellites operate at a very high altitude above the Earth, where the bending of space-time is less. Time at these altitudes is faster than on Earth due to the effects of General Relativity. Second, the high relative speed between the GPS satellites and GPS devices on Earth has the effect of making satellite clocks appear to run slow compared to the clocks on Earth *("moving clocks run slow")*. These

10 The second detection was reported by Jennifer Chu in *MIT News* in June 2016.

effects are in opposite directions. Both must be accounted for in order for GPS to work.

It turns out that the effect of the altitude of the GPS satellite clocks due to General Relativity is much greater than that of the relative motion between the satellites and Earth-bound GPS devices. This larger effect of altitude on the GPS satellite clocks is corrected when the satellite is launched by setting the clocks to run appropriately slow to overcome the faster time passage for the high altitude satellites. This allows the satellite clocks to run at the same speed as the clocks on Earth. The individual GPS devices on the ground make the corrections for the smaller Special Relativity effect of the relative motion. The GPS satellites send information about their speed and direction to Earth-bound GPS devices, so the appropriate time adjustments can be made. This is one real world example of the application of relativity in making our lives better. Figure 11-11 shows the faster time experienced on the GPS satellites that had to be corrected by setting the satellite clocks to run slower.

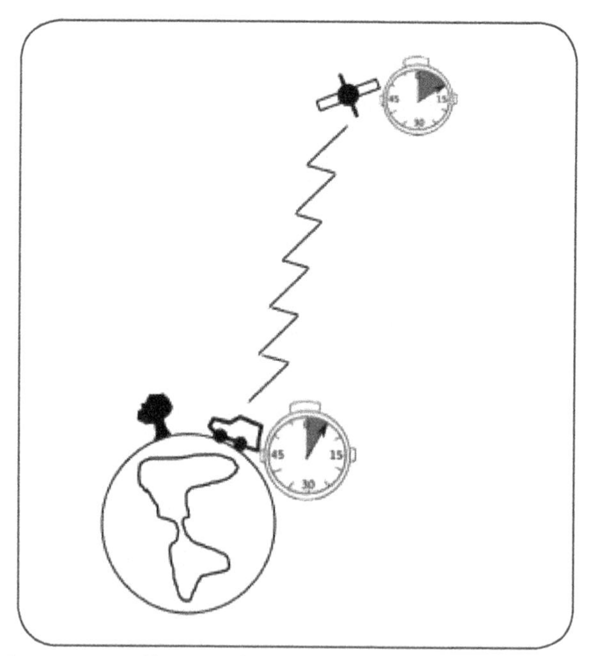

Figure 11-11. Modern GPS satellites apply corrections to their clocks to account for the less compressed space-time at a higher altitude. *Without correcting for the faster time at high altitude, today's GPS would eventually be off by miles. The GPS devices on the ground correct for the smaller effect of relative motion between the satellites and the Earth-bound GPS receivers. These corrections are made using information sent to Earth from the satellite.*

Chapter 12
Science and Religion

This chapter addresses the following topics and questions:

12.1 The goal of science. *What is it that scientists do? How do their goals relate to Einstein's discoveries about relativity?*

12.2 Science and religion. *How can our growing understanding of science inform our religious beliefs? How can scientific exploration enhance and even inspire our religious experience? How can religious experience inform and guide science? What can and should religions do to keep up with science?*

12.3 God's DNA. *What can we conclude about God from what we now know about the Universe?*

12.4 Beyond DNA. *How have humans taken evolution to new levels? How has the evolution of information expanded beyond what can be encoded in biological beings?*

12.5 The Universe and God. *What questions about Creation and God do emerging scientific discoveries raise? How can we begin to answer these questions?*

12.6 Closing thoughts. *What are the main takeaways of this book?*

Einstein offered a new way of understanding the Universe that now stands as the foundation for much of physics. His work illustrates science at its best. It also raises questions, as he did, about the relation between science and religion. This chapter discusses these issues.

12.1 The goal of science. Science is a process of developing an understanding of the nature and behavior of the physical and natural world. There is no better example of how it works than the one provided by Einstein.

The Universe is not always what it seems. It is what it is. It is the scientist's job to figure out what the Universe is; how the Universe behaves; and why it works the way it does. To do this, it is sometimes necessary to set aside prior conventions and pre-existing notions about how we think nature works. It requires developing an unbiased, deep understanding of the laws of nature, sometimes putting aside established dogmas of all types — scientific, religious and cultural. It means exploring natural laws in an unbiased way; building on what went before, but sometimes looking for novel solutions, in the form of hypotheses, when traditional explanations fail to work under all or newly emerging conditions. It then requires putting the new hypotheses to the test by subjecting them to verification by careful measurement and scientific experimentation.

Einstein set the bar high in this regard. He saw the unexplained phenomenon of an unvarying speed of light, and the confusion of distinguishing between acceleration and gravity, and was able to develop entirely new hypotheses that could explain these conundrums. He was able to move beyond conventional thinking and derive explanations that made sense and worked, even though they required a completely new way of thinking about time, space and geometry. When the scientific community pushed back and questioned Einstein's solutions, because they ran counter to common wisdom and scientific dogma of the time, Einstein dug deeper. He did the math. He determined the relationships. He showed how his new theory was able to account for the previously unexplained anomalies of the orbit of Mercury. And then he identified additional measurements and experiments that could further demonstrate the scientific beauty and robustness of his conclusions.

His persistence paid off. The measurements he called for, and subsequent experiments, showed his theories of relativity to be correct. They accurately described the behavior of light, gravity and time in a way that other hypotheses and earlier theories could not. Relativity today is a broadly accepted theory about how the Universe works. In fact, it is the basis for many new technologies that now contribute to our lives every day; technologies that would not be possible without an understanding of relativity.

But Einstein himself was not immune to the temptation of dogmatic thinking. When the simplest formulation of the General Theory of Relativity predicted that the Universe had to be either expanding or contracting, Einstein felt that something had to be wrong because he, like many physicists at the time, believed that the Universe was static – neither expanding nor contracting. There was no evidence indicating otherwise at the time. And there was a place in the General Relativity equations for adding a term that would determine the amount of expansion or contraction. Einstein carefully selected a value for this term that would define a Universe that was static, not expanding or contracting. He called this term the "cosmological constant." A few years later, other physicists carefully solved Einstein's original equations and found that they actually predicted an expanding Universe. At about this same time, astronomer Edwin Hubble discovered that, indeed, the Universe was expanding as the equations, when properly solved, predicted. Einstein later said that selecting a value for the cosmological constant was the biggest mistake of his career and removed the added term.[1]

In hindsight, some might point to Einstein's reluctance to fully embrace new interpretations of Quantum Theory, a theory he helped develop, as dogmatic thinking. However, this would be unfair and untrue since during Einstein's lifetime, the new Quantum Theory's predictions were untestable and required, or at least implied, faster-than-light communication! Einstein played an important and central role in debating the emerging interpretations of Quantum Theory. He, along with other physicists, offered numerous "thought experiments" that analytically explored the merits and shortcomings of the theory's predictions, and argued that the new way of understanding quantum mechanics being proposed could not be true because they required what

1 Ironically, today's astronomers have discovered that the Universe is not only expanding. It is expanding at an accelerating rate due to a repulsive force distributed throughout the Universe. This force is still not understood, but its effect on the expansion has been measured. Einstein's cosmological constant is being used to account for the repulsive force and accelerating expansion. The term may be needed after all; just not the way Einstein originally envisioned it being used.

Einstein called "spooky action at a distance." It was actually Einstein who first pointed out that if the proposed interpretations were true, "entangled" subatomic particles would immediately have to adopt correlated descriptive parameter values when those values were measured to exist in the correlated (entangled) particle, no matter how far away entangled particle may have moved. That is, the entangled particle would have to immediately know when the other entangled particle was measured so it could immediately assume those correlated parameter values. This is spooky action at a distance, indeed.

Today, the probability-based Quantum Theory is widely accepted science, in spite of the faster-than-light communications that it seems to imply. In fact, experiments have been performed repeatedly to demonstrate the strange associations between "entangled particles" across relatively large spatial distances — even as far as a few hundreds of miles! The mechanism that enables the seemingly faster-than-light communication is still not understood.[2]

Today, both Einstein's theories of relativity and Quantum Theory are considered accepted science. While the two theories still don't work well together, they both provide ways of understanding different aspects of the Universe that are borne out by experiment. On very large scales involving, for example, stars, planets, and galaxies, relativity is the accepted theory able to explain the measured phenomena. At subatomic scales, it is Quantum Theory that rules. During his lifetime, recognizing the incompatibility between relativity and the new Quantum Theory, Einstein worked with other scientists to find solutions that could combine these two theories into one "Grand Unified Theory." This work continues, but so far, one has not been found.

Einstein's theories of relativity represented a major breakthrough in science and today stand as some of the greatest scientific discoveries of all time. It was Einstein's willingness to set aside scientific beliefs of the day and explore a new way of seeing and understanding the Universe that made the breakthroughs possible. It stands as testament to the scientific method in which scientific consensus is formed through a process of experimentation and scientific evidence gathering. It stands as a model for how science should work. The process continues and theories that even go beyond relativity are possible. It is even possible that a "Grand Unified Theory" that

2 See Brian Greene's book, *Fabric of the Cosmos: Space, Time, and the Texture of Reality* for a discussion.

combines General Relativity with Quantum Theory, that is able to describe the Universe across all size scales, may one day be developed.

12.2 Science and religion. Science and religion both focus on understanding our place in the Universe. But they arrive at answers that differ in their nature and certainty. While each applies fundamentally different perspectives and methods, science and religion can compliment and inform each other.

Both science and religion are concerned with questions about our place in the Universe, and its creation, but they come to these questions from different perspectives and use different methods. Science addresses these questions using the scientific method, already summarized, in which observations generate hypotheses that are subjected to rigorous testing. Religion, on the other hand, comes from human spiritual experience and exploration. It tries to answer existential questions of human and world purpose, and to relate the answers and insights gained to define societal values and norms of moral behavior, and to inform ethical decisions within society. Religion also addresses the question of whether there is a God, and tries to discern the nature of God, if one exists. These are questions that cannot be settled scientifically. When religious answers go beyond what can be determined objectively or scientifically, they draw from human feelings and inspiration; they come from a spiritual source.

Both science and religion can guide, inform and help satisfy our search for truth and meaning. Both have a vital role in the human quest to understand existence and the human condition. Religion applies spiritual insights to explain things that are unknown. It builds on human feelings and emotions to establish connections between current human experience and the larger Creation that is beyond our ability to understand, and even fully appreciate. It reminds us that even as science expands our understanding of the workings of the Universe, there will still be questions of purpose and meaning that will remain in the realm of religion.

Pope Francis put it this way:

> *"...faith encourages the scientist to remain constantly open to reality in all its inexhaustible richness. Faith awakens the critical sense by*

preventing research from being satisfied with its own formulae and helps it to realize that nature is always greater. By stimulating wonder before the profound mystery of Creation, faith broadens the horizons of reason to shed greater light on the world which discloses itself to scientific investigation."[3]

While religion offers spiritual explanations for the unknown, science provides an objective understanding about the scale, power, nature and workings of the Universe that informs and guides larger questions about ultimate truths being pursued by religion. Both work and grow together and contribute to our understanding of Creation. For example, scientific discoveries about the scale and grandeur of the Universe put religious Creation stories in a grand new context. They inspire, or at least should inspire, the building of new religious stories, myths and teachings that help us relate to the power and mystery of Creation in the context of our growing objective scientific foundation. Both science and religion play important roles in our search for ultimate answers.

Einstein put it this way:

"Now, even though the realms of religion and science themselves are clearly marked off from each other, nevertheless there exist between the two strong reciprocal relationships and dependencies. Though religion may be that which determines the goal, it has, nevertheless, learned from science, in the broadest sense, what means will contribute to the attainment of the goals it has set up. But science can only be created by those who are thoroughly imbued with the aspiration toward truth and understanding. This source of feeling, however, springs from the realm of religion. To this there always belongs the faith in the possibility that the regulations valid for the world of existence are rational, that is, comprehensible to reason. I cannot conceive a genuine scientist without that profound faith. The situation may be expressed in the image: Science without religion is lame and religion without science is blind."[4]

3 *The Light of Faith, Lumen Fidei. Encyclical Letter.* Lumen Fidei of the Supreme Pontif Francis to the Bishops, Priests and Deacons, Consecrated Persons and the Lay Faithful on Faith.

4 Einstein, Albert. Essay: "Science and Religion." In *Ideas and Opinions.* by Albert Einstein. Pages 45-46. (also available at http://www.sacred-texts.com/aor/einstein/einsci.htm)

Today, science and technology are pushing our knowledge of the Universe to new, unimagined levels, making the relation between science and religion ever more important. All religions have addressed the question of Creation. Religious stories describe how the world was created from the perspective and understanding of the times when they were created. For example, religious Creation stories, and teachings more generally, for agrarian societies were typically very Earth-centered, reflecting on the cycles and rhythms that sustain and nourish life on Earth. The real scale of God's Creation, revealed by recent scientific discoveries, could not have been imagined, let alone understood, in ancient times.

The work of Einstein and other scientists has done much to inform and influence our beliefs. For example, we now know that our Universe started 13.8 billion years ago in a huge creative event called the Big Bang from which the entire Universe came into existence, and then evolved to its present state, through various processes in which the material that forms the Earth and life were formed. We know that we live in a huge galaxy called the Milky Way that contains over 100 billion stars and stretches over 100,000 light-years across, and that many of the stars in the Milky Way have orbiting planets like our Sun does. We also know that there are over 100 billion other galaxies. And we know that the Universe is continuing to expand at an increasing rate and that space and time do not exist separately, but only in combination as an integrated entity that Einstein called space-time. Closer to home, science is telling us that our human activities are changing Earth's climate with potentially serious near-term consequences and with possible long-term catastrophic impacts.

This scientific knowledge does not diminish our spiritual and religious beliefs and needs; it enhances them. It expands our religious perspective and allows us to ask religious questions with a fuller appreciation of the Universe and the unimaginable scale of Creation. It informs, but does not answer the larger religious question of Creation's purpose. Why was the Universe created? What is our role in it? What is the human relationship with other parts of the Universe? What is, or should be, our relationship with other life forms on Earth? What is the purpose of human life? What is the purpose of the Universe? How should we live in accordance with that purpose? How should answers to these questions guide our personal goals and lifestyles? What is our responsibility to the Earth? How should we respond to the threat of climate change? These are religious and spiritual questions that are informed and guided by modern science.

And religious questions provide inspiration and direction to the conduct of science, and for a more meaningful life in general. It was religious questions, along with a driving scientific curiosity, that inspired Einstein and other scientists to probe the deepest and most meaningful scientific questions. In fact, these are religious, as well as scientific questions. The scientific answers, like those provided by Einstein, informed our understanding of Creation, while leaving the ultimate religious questions in place to be contemplated in a grander and more complete context. In other areas of science, like biology, psychology or medicine, religious beliefs and questions can help us answer ethical questions and guide us in setting directions and limits on scientific endeavors. For example, if our religious beliefs are that humans have some responsibility as stewards of the Earth, this can guide us in our individual lifestyle choices, as well as inform our societal decisions and responses to problems that are encountered, such as environmental pollution and possible human causes to changes in Earth's climate.

The recent book titled, "The New Universe and the Human Future" by Nancy Ellen Abrams and Joel Primack provides a thorough review of our current understanding of the Universe from cosmological and physical scientific perspectives, and recommends that current religions integrate the latest science into their religious beliefs. They call for building a common scientific understanding of the Universe that we all can agree upon, and from which all religions can build, within the various religious and cultural settings of the world.

This is not to say that current religious myths need to be discarded or changed. On the contrary, they need to be understood in the context of an evolving foundation of science. Rather than argue about which ancient story is literally true, religions should understand their myths and stories as historical foundations and inspired writings that can inform modern religions in a way that also reflects the latest scientific knowledge. The goal is to seek the deepest truths represented rather than simply explore literal interpretations. This is well captured by a Native American tribal chief who is attributed to have said, when telling his tribe's religious stories, "I don't know if this actually happened, but I know that it is true." This changes the focus from the words to the underlying meanings represented. The goal is to understand our religious stories in ways that can inform, inspire and guide us in our lives today.

To summarize, the world's religions can be the bridge between emerging scientific knowledge and personal religious beliefs. Both science and religions grow by informing each other to build deeper and more meaningful truths based on a growing understanding of the Universe and Creation. Old religious myths take on new deeper meanings as scientific foundations are strengthened and expanded. For example, the Pope's recent encyclical on climate change draws from long-standing Catholic religious beliefs to guide and inform societal responses to the modern threat of climate change. In some cases, new religious myths will be created as contemporary religious leaders tell new stories that relate old myths and emerging science to modern times; new stories that answer religious questions in ways that are relevant to the latest science and modern life; stories that everyone can believe in regardless of their literal truth, or lack thereof.

12.3 God's DNA. Relativity and other recent scientific discoveries have given us new insights and understanding about how the Universe was formed and how it has evolved, but still don't answer the ultimate God question. This is still a matter of belief and religious theology. This section discusses how the latest understanding of science can inform and influence our religious quest, while not replacing it. Our scientific efforts are discovering God's blueprint for Creation; God's DNA that we share. This understanding can inform our religious beliefs.

The book of Genesis says "Let us make man in our image, after our likeness..." (Gen 1:26). If you accept the premise that God exists, and the Bible is an expression of God's purpose, then this is one of the few passages in the Bible that is actually, literally true, at least when understood in the context of modern science. This is discussed in this section.

First some background. We live at a time that is historically unprecedented. We are the first generation to know (approximately) when the Universe started and to have a general understanding of how it evolved to its present state. This knowledge is at a coarse level, lacking in many details. Nevertheless, our scientific knowledge today provides an objective appreciation of the scale and age of the Universe. Of course science still doesn't explain what caused the Universe to be created or why. Does the Universe have a purpose? If so, what is it? And what is the human purpose within it? These questions are the domain of religious inquiry and belief.

Nevertheless, we know that the Universe started 13.8 billion years ago at a single point in space-time according to scientific calculations. It began with a huge explosion-like beginning that scientists call the Big Bang. There is convincing evidence that during this beginning, space-time expanded dramatically (to say the very least), growing quickly in size and creating more and more space-time as it grew. Scientists believe that the early expansion was so rapid that in the earliest stages, it expanded at speeds exceeding the speed of light, perhaps at a time before Einstein's theories of relativity were created or were operative.[5]

At the very beginning, the Universe was pure energy expanding into newly created space-time. During the first second, scientists believe elementary (subatomic) particles of matter began to form. From these particles, the forces of gravity, electromagnetism, and the strong and weak nuclear forces that hold atoms together developed. More familiar particles we know as protons, neutrons and electrons formed next. Again, this was all during the first second after the Big Bang, after the creation of the Universe. During the first few minutes of the Universe, the first and simplest elements (hydrogen and helium) were formed. As the Universe expanded, the hydrogen and helium atoms came together under the force of gravity to form stars. The early stars were much larger than our Sun. They converted the hydrogen and helium into heavier elements, such as carbon, oxygen, and nitrogen, like stars do today. The Universe was large enough, and transparent enough, to allow the first light to shine after just 400 million years. At this time, for the first time, energetic photons traveled through space-time as they do today, moving at speed "c" following the gravity-influenced contours as described by Einstein. If God said, "Let there be light," they said it about 400 million years after they created the Universe.

As the Universe grew, more stars were created within the many galaxies that were formed. When massive stars used up their nuclear fuel, they sometimes ended their lifespan in cataclysmic explosions called supernovas. Like supernovas today, these powerful explosions forced subatomic particles into new larger, heavier elements. Not just carbon and oxygen, but all the

5 The NASA website has an excellent discussion of how the Universe started and evolved. Some of material for this section was drawn from it: www.nasa.gov/vision/universe/starsgalaxies/wmap_pol.html.

elements we see today, like silicon, gold, silver, uranium, and so forth. These larger, heavier, more complex elements later became part of second and third generation stars and planets, like the Sun and Earth. The elements created in stars and supernova eventually combined to form molecules that make up things like water and carbon dioxide that we find on Earth, and other planets and moons in the solar system, and throughout the Universe.

As planets and moons were formed, even more complex molecules evolved. This chemical evolutionary process continued as more and more complex chemical structures formed, like hydrocarbons and acids. The increasing complexity reflects a growth of information in the form of laws of physics and chemistry. This information growth ultimately resulted in the forming of very complex molecules, such as amino acids, that became the basis for life. These amino acids eventually evolved into molecules of DNA (deoxyribonucleic acid), which is the foundation for life on Earth (and perhaps elsewhere). That is to say, the Universe evolved to form life! This is shown in Figure 12-1.

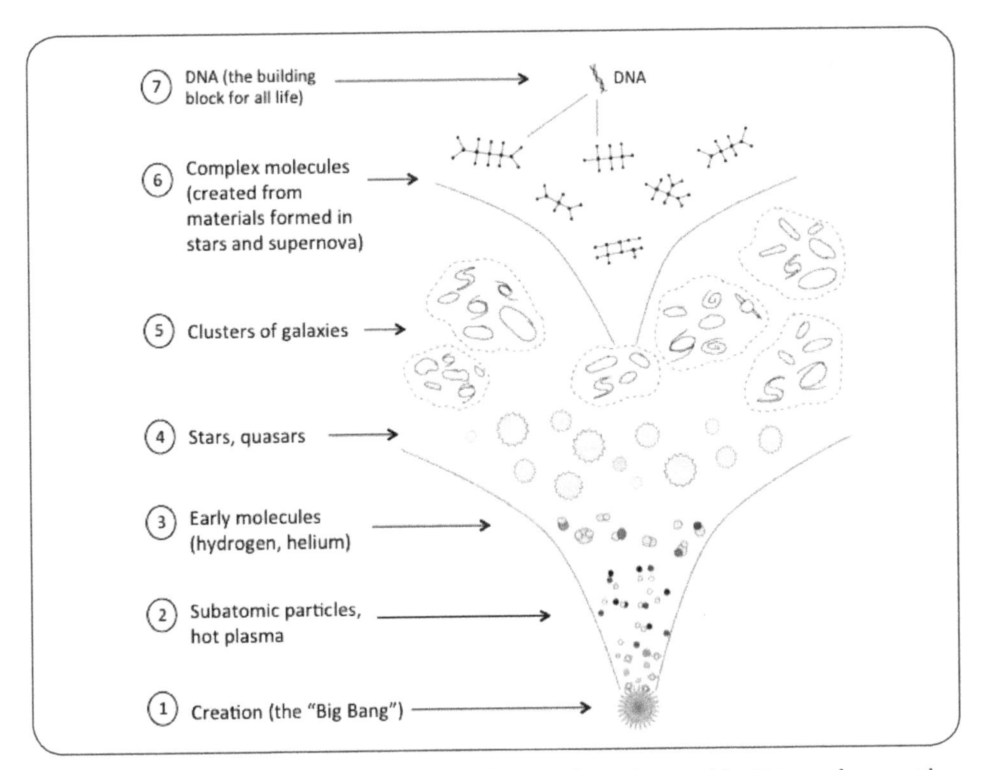

Figure 12-1. The early stages of Universe evolution. *The evolution of the Universe began with the Big Bang and evolved over the subsequent billions of years to create many forms of matter and energy, and ultimately self-replicating molecules that formed the basis for early life.*

DNA is a unique molecule that is critical for enabling life because: (1) it is able to encode large amounts of information – information that defines and regulates the form and functions of life; and (2) it can make replicas of itself, thus passing that information to new DNA molecules that are created. This allows the information encoded in the DNA molecule to be passed to subsequent generations, with small changes reflecting the sharing of the information between different individuals, and occasionally larger changes called mutations that appear to be random.

The DNA molecules form the basis for all life on Earth. They reside in every living cell and contain the information that specifies and regulates the functions of every living being. Just as the DNA in flower seeds has the information to tell the new flower how to grow, the DNA in humans has the information that specifies everything about us from eye color, to the location, structure and functioning of all our body's organs and tissues.

It also contains the instructions our bodies use to manage growth from baby to an adult. Because new DNA is formed when two people come together to share their respective DNA, the baby's DNA draws a subset of information from each parent. The new baby inherits some information (and traits) from each parent. Over many generations, the process allows the species to gradually change (to evolve), making small changes over time; changes that may or may not improve the ability of a new generation to survive. Some changes can improve the ability of subsequent generations to adapt to a changing climate or to new diseases, while others may reduce the ability to survive. As offspring that are better adapted for survival thrive, and those less well adapted are more likely to die, the evolving DNA forms the mechanism supporting the process of adaptive evolution; the process of life selectively adapting in ways that facilitate long-term survival and long-term compatibility with changing environments. It's nature's way of making sure everything works together in an interdependent, self-supporting whole; an interdependent web of existence.

As the DNA changes from generation to generation, either due to normal mixing of DNA during reproduction, or especially from larger abnormal mutations, the functions of the cells and larger body parts that they define can also change. Organs within organisms can become more specialized or even change over time to perform completely new functions. The organisms themselves can evolve, becoming more complex and more intelligent.

Through this DNA-enabled evolutionary process, simple early life forms have evolved to form the wide range of plants and animals we find on Earth today. Humans are just a recent and more complex manifestation growing from the process. While our DNA is the same basic molecule as the DNA in other life forms (plant and animal alike), it is more evolved and enables more complex functions such as the cognitive abilities we enjoy. And the process is continuing.

If the Bible is correct that God created man (and presumably woman) in the image of God, this was done using DNA. It is our DNA that defines our appearance. It determines our eye color, the shape and size of our nose, the color of our skin and hair, the shape of our face, how tall we are, and everything about us. We appear the way we do because of our DNA. If we were made in God's image, it is because we share God's DNA.

But DNA is the biological blueprint for all life, evolved from the very first life form that appeared on Earth. Our DNA is the genetic offspring of this first DNA. This is true for all life on Earth, including both plant and animal life. This is shown in Figure 12-2. If humans are created in God's image, and share God's DNA, then it is reasonable to conclude that all life is an expression of God's image through our related DNA.

But our DNA not only connects us to all life on Earth, it also connects us to the larger, evolving Universe, since life grew out of the larger process that created the Universe, as described earlier and shown in Figure 12-1. Whatever it was that created the Universe, it started a process that led to all life forms, as well as to the entire Universe. If there is a God, or if the Universe can be thought of as God, then it is a God that is connected through the laws of nature (e.g., the laws of physics and chemistry), defined by characteristics encoded through the laws of nature and reflected in the DNA of the many evolving life forms we see. In this view, we are not only created in God's image, we are an integral part of God and God's evolving image. We share God's DNA. And if God is the Universe, or is even just expressed through the Universe, it is a Universe connected and defined by an evolutionary process that connects all of Creation to the initial beginning, the Big Bang. The Universe is an expression of the creative force unleashed at the instant of the Big Bang; an expression that continues to unfold here on Earth and everywhere throughout the Universe.

In addition to being connected by the laws of nature, and the evolving knowledge that it encodes, the Universe is also connected in real time by the various forms of field radiation, as science is discovering. For example, electromagnetic waves, like gravitational fields, are everywhere throughout the Universe providing us with a detailed view of the larger Creation, past and present. These radiation fields are available for informing life everywhere about the nature of the Universe and about Creation itself. They carry the information that defines and informs the Universe about itself; they allow us to discover how the Universe works, how it started and how it is evolving.

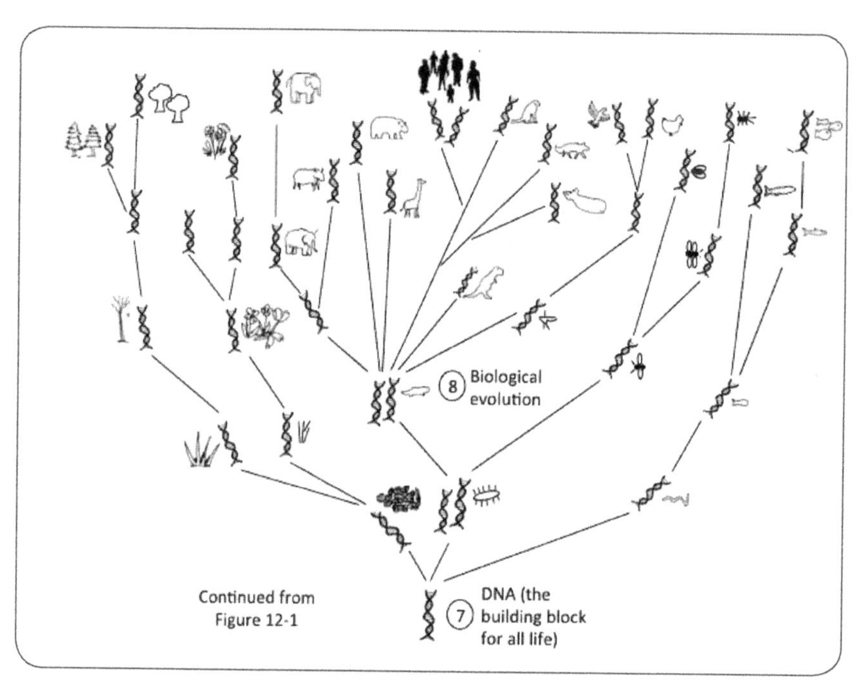

Figure 12-2. The evolution of life. *Recent periods of evolution of the Universe builds on self-replicating molecules called DNA. The DNA formed the basis for early life and allowed it to evolve into a wide range of life forms, all related by common ancestral DNA, and all having the appearance and functions defined by the DNA*

12.4 Beyond DNA. Humans play a unique role in the evolution of the Universe. As biological information has evolved to become more complex, especially within the DNA molecule, the most complex life form that we know of (humans) has developed intelligence capable of creating and evolving information outside of the biological realm.

We as humans have reached a pivotal point in the evolutionary process. With the evolution of our large brains and specialized hands that can grasp and manipulate things, we are able to develop information and intelligence outside of our biological bodies. For example, we have been able to build computers that can store large amounts of information of all types (e.g., pictures, data, music and specifications) and process that information much faster and more efficiently than we can with just our biological tools (our brains). We are also able to build scientific instruments, such as sensors and equipment that can detect and measure things beyond the capability of our five human senses. We have built spaceships and robots that can carry our sensors to remote locations beyond Earth to seek out new information from outside our immediate environment. We have used our telescopes to peer into the far reaches of the cosmos. We are also able to use our instruments to look inside the molecules and atoms that form the matter that make up the Universe.

And we have been able to use this information, and the tools we have developed, to figure out how the Universe works, as Einstein did. We have even been able to discern the historical origins of the Universe drawing on the many clues the process left for us to discover. And we have discovered the secrets of DNA, the blueprint for life and the engine of evolution. We have been successful at all these endeavors because we have been able to accumulate, store and manipulate knowledge outside of our biological bodies; and have been able to use tools and sensors that expand our own limited sensing abilities. This has allowed detailed exploration of our world and the Universe. Because we are a part of the Universe, it is a process of the Universe becoming conscious of itself, summarized in Figure 12-3.

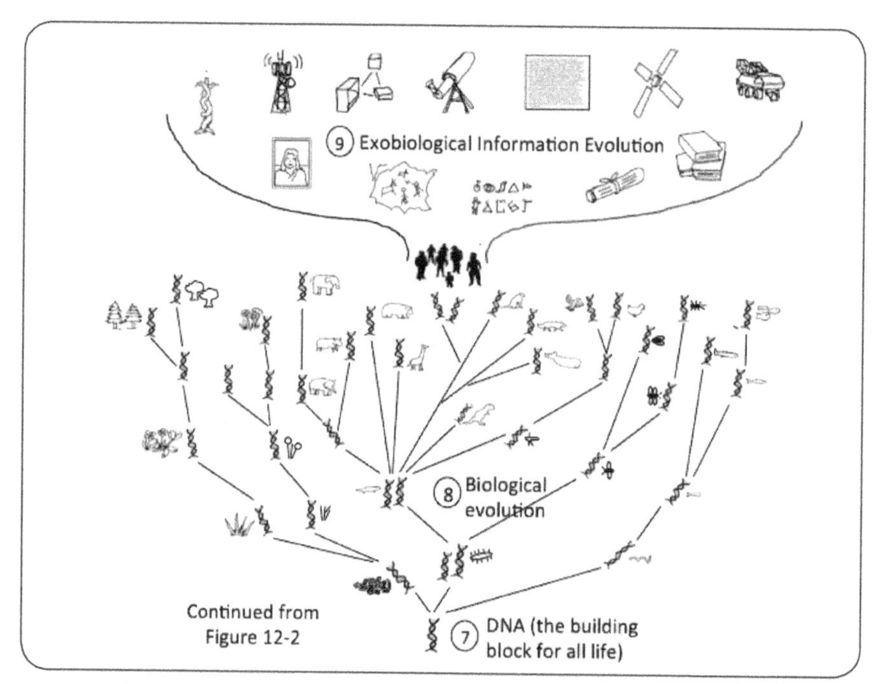

Figure 12-3. The evolution of information beyond biology. *Human beings are unique among life on Earth in the ability to develop information outside our bodies and technology capable of working beyond our biological capabilities.*

12.5 The Universe and God.

Our ponderings about relativity and the evolution of the Universe ultimately raise questions about the existence and nature of God. This section discusses some of the religious questions that are raised and what answers might be offered.

As we use our instruments and tools to discover the nature and workings of the Universe, it continues to raise the ultimate questions: what caused the Big Bang? What made the Universe grow and evolve the way it did? Was there a blueprint for the overall evolution "baked-in" at the beginning or was the process random? Why was the Universe created? What was there before the Universe was created? Of course, these are mostly religious questions. It is not possible to answer them objectively and definitively – at least not with our current technology and level of scientific understanding. Through science, we can determine with greater and greater accuracy how the Universe works, how it has evolved, and maybe even what initiated the Big

Bang. But the questions of why it started, its purpose and our purpose within it remain questions of faith and religion.

All religions have tried to provide answers to these questions of Creation and human purpose. The religious answers are usually based on the experience and understanding of the society and culture involved. For example, agrarian societies saw cycles of nature reflected in the changing seasons and cycles of life that surrounded them; they formed religious understandings and myths based on that understanding. For example stories about the origin of corn and about droughts and floods that disrupted the normal cycles of nature, were made part of the religious traditions. When early societies appreciated and understood the vital role of the Sun in nourishing the Earth and its food sources, religions that worshiped the Sun and its cycles became predominant. As more sophisticated social structures were formed, religions that tried to understand and strengthen those structures were developed. Stories of human conflict and suffering outside of "Eden," and rules for guiding family and community activities formed. Stories of cooperation and brotherhood were included within the religious fabric, and helped hold society together.

Today, we have a very complex, multifaceted society that is merging many cultures together and is bringing new, unprecedented scientific knowledge about the larger Universe to bear. This calls for an evolution of our religions in a way that can address and support these complexities. We know that the early Universe created the initial atomic particles that became the elements on the periodic table that evolved into complex molecules that were able to encode information. We know that we humans, and our religions and scientific knowledge, came out of this process. And we know that cultural change, and the growth of scientific knowledge, is developing with unprecedented rapidity.

We need our religions and religious institutions to adapt and respond to these new fast emerging cultural and scientific changes. The old stories need to take on new meaning, and new stories and new religious support structures need to be developed. Whatever the source of the Universe was, and why it was created, it is clear that the Universe itself is evolving and that we are part of that evolution. We came out of the Universe and from its larger evolutionary process. We are an integral part of the Universe. We were not created separately and put in it. We need to incorporate this knowledge into our religious stories and evolve our religious teachings and traditions to reflect this new understanding. We need to develop new ways to understand our old religious myths and stories in the context of this evolving reality, and form new stories that expand on them. And our religious leaders need to help guide this transition and personal growth.

There will always be a need for religions to seek answers to the God and Creation questions — is there a God, and if so, what is its nature? For many, God is a word to describe that which we can't understand, but which has profound creative power and presence. For others, God will continue to assume a more personal, nurturing nature. Still others will take an agnostic or atheistic view. There will not be a single answer that will work for everyone or for all religions. It is important to respect all views on the ultimate questions about God and Creation, since there are no objective answers.

Sometimes I think of the Universe as God; a growing connected being or intelligent entity of which we are an integral part. But of course, something created the Universe, so maybe we should call the creative force that started the Universe God. Does God exist outside of the Universe? Beyond the Universe? Is God a field? Is God quantum energy? Is God a being? Does God communicate through quantum energy fields? Can we connect to the God field — to the "Force" — through our attention? Through meditation? Prayer? Scientific and academic study? Deep skills mastery? *What are thoughts? Is there a God?* We aren't sure.

We are like a cell in the stomach trying to figure out its own nature and understand the larger context of its existence. The stomach cell sees intermittent flows of material rich in sugars and proteins flowing in. It sees chemicals in the stomach working with various bacteria to breakdown that material which is taken into the cells and combined with oxygen from elsewhere to generate energy. The cells may notice the remaining material flushed downstream for elimination. Eventually the stomach cells sense the large overall movements of the larger existence in which the stomach is located. Based on this experience the stomach cells "figure out" that they are part of something bigger, something that moves and has cycles of activity and rest. Blood flow is increased during the periods of activity, while stillness and quiet characterizes the less active periods. The cells come to realize that they are part of a larger being of some type. As they gather more and more information, they may even "surmise" that they are part of God's Creation and play an important role in the workings of that Creation. They develop religious beliefs that represent this incomplete understanding. They know that they are part of something bigger that they call God, or call God's Creation.

We are in a similar situation, but in the larger Universe. We started out seeing the Sun, Moon and stars circling the Earth and believed that was what they were doing. Today, our view is wonderfully broad. We know that the Earth and all the planets in our solar system orbit the Sun and that the Sun is moving in a circular fashion around the black hole in the center of our galaxy that we call the Milky Way. We know that there are planets orbiting many stars in the Milky Way and that there are billions of other galaxies scattered across the Universe. We understand these things about how the Universe is today, but also have knowledge about how it has evolved over time; a Universe that is billions of light-years across and that was created 13.8 billion years ago. But, like the stomach cells, we are becoming aware that we are part of something bigger; that we are part of an intelligent evolving Universe, yet can't imagine with any certainty what might be outside, how or why it started, or its ultimate purpose. These remain religious queries.

When you boil down what we know, we know this: the Universe is made of information; information expressed through the laws of physics and chemistry, and encoded in the many forms of matter, energy, life and behavior. The expression of the information is evolving, becoming ever more complex and complete. On Earth this is expressed in the DNA molecule that encodes the information enabling humans and other life forms to live, reproduce and evolve. And the information now exists and grows in external structures created by humans outside of the biological realm. As life and information evolve in this way, the Universe becomes more complete and more aware of itself.

We are part of nature's evolution, a process of nature becoming self-aware; discovering its own nature. It seems our role (responsibility), at a minimum, is to pass along our information (DNA and externally derived learning) to following generations; and to make sure there is a healthy, living planet for those generations to live and evolve on. All religions and religious leaders, regardless of their specific beliefs about God, need to call on all people to be stewards of the Earth and to accept our immense responsibility to future generations. Pope Francis powerfully expresses this message in his recent *Encyclical on Climate Change*, as follows:

"I urgently appeal, then, for a new dialog about how we are shaping the future of our planet. We need a conversation, which includes everyone since the environmental challenge we are undergoing, and its human roots, concern and affect us all...

"The climate is a common good belonging to all and meant for all. At the global level, it is a complex system linked to many of the essential conditions for human life. A very solid consensus indicates that we are presently witnessing a disturbing warming of the climatic system. In recent decades, this warming has been accompanied by a constant rise in sea level and, it would appear, by an increase of extreme weather events, even if a scientifically determinable cause cannot be assigned to each particular phenomenon. Humanity is called to recognize the need for changes of lifestyle, production and consumption, in order to combat this warming or at least the human causes, which produce or aggravate it...

"This necessarily entails reflection and debate about the conditions required for life and survival of society, and the honesty needed to question certain models of development and consumption..."

Our concern must be to preserve and promote life on Earth over the longest-term; our focus must be the health of humans, all life and the Earth for millenniums to come. We should all be able to agree on this! At a minimum our individual and collective decisions and actions should consider ramifications over at least the next 100 to 500 years, a timeframe our children, grandchildren and great-grandchildren will inherit. This is a responsibility we share with all people and all religions.

12.6 Closing Thoughts. This book was written to introduce the concepts of relativity and to make them easier to visualize and understand. It concluded with an exploration of the relationship of science and religion, as typified by the discoveries of relativity. This exploration included consideration of the ultimate questions of Creation that are raised when science is inspired by religion, and religion is informed by science.

This book focused on describing Einstein's theories of relativity. Like many books on relativity, it uses simple two-dimensional illustrations to help us see how relativity works. While these illustrations help us visualize the concepts of relativity, they do not completely and accurately capture the underlying science. This is because Special and General Relativity are based on a different kind of geometry and described by very complicated mathematics. Nevertheless, the illustrations help us to appreciate the general principles and conceptual workings of relativity from perspectives we familiar with.

This book also described how Einstein's accomplishments modeled the way science progresses when it works properly. It described how the development of relativity demonstrated the process of forming hypotheses based on observations available, and then defining tests that could demonstrate the correctness of those hypotheses. The theories that Einstein developed have been subjected to extensive verification testing and now form an accepted foundation for modern physics.

Finally, this book explored the relationship between science and religion. It described how science and religion differ in their nature but compliment and inform each other in practice. It described how science and religion can work together as part of the Universe's struggle to become self-aware; and toward the fulfillment of the human role and purpose in that struggle.

The book was designed to gradually build an understanding of relativity through a continuing pondering process; pondering about how relativity works and what relativity says about how the Universe is made. To this end, the reader is encouraged to re-examine the figures in the book, review the summary in Table 1 and do the exercises in Appendix C.

I hope you enjoy pondering this book.

Appendix A
Simple Derivation of Special Relativity Using the Pythagorean Theorem

For those interested, this appendix addresses the following question:

A.1 Simple calculation of space compression and time slowing. *Is there an easy way of seeing how the Special Relativity mathematics works to slow time and compress space on moving objects?*

A.1 Simple calculation of space compression and time slowing.

This section presents a simple way of seeing how the mathematics behind Special Relativity works.

This appendix uses the Pythagorean theorem to derive formulas for determining the amount of space contraction and time slowing experienced when applying the principles of Special Relativity. It is based on the analysis of Richard Wolfson in his book "Simply Einstein."

A quick review of high school algebra will recall the Pythagorean theorem as a simple way to determine the length of the sides of a right triangle. To review, the theorem says that for any right triangle the hypotenuse squared is equal to the sum of the squares of the other two sides ($c^2 = a^2 + b^2$). This allows us to find the hypotenuse (c) by taking the square root of the squares of two shorter sides added together ($c = \sqrt{a^2 + b^2}$). Figure A-1 depicts a right triangle with sides a, b, and c.

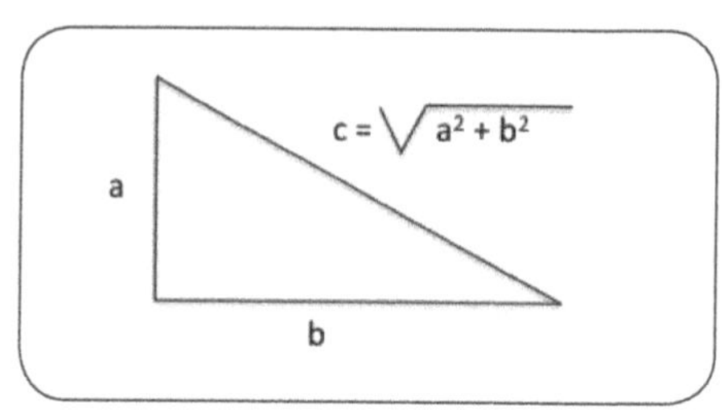

Figure A-1. The Pythagorean theorem. *The Pythagorean theorem defines the length of the hypotenuse of a right triangle in terms of the lengths of the other two sides.*

Also recall that distance traveled is simply speed times time *(distance = speed × time)*. For example, if you drive for two hours at 50 miles per hour, you will travel 100 miles *(2 hours × 50 miles per hour = 100 miles)*.

Now recall the time slowing effects of a spaceship moving at near the speed of light. Also, recall the operation of a light-clock described in Chapter 5. We will apply the Pythagorean theorem to understand how time slowing and space compression works using a light-clock moving at near the speed of light from left to right. This is shown in Figure A-2. The length of the light clock (i.e., the distance the light travels in one direction inside the clock) is given as L. The path of the light for one clock tick (light goes to bottom and returns to top) is 2L when the light clock is at rest. When the light-clock is moving, the length of the light path as seen from a stationary observer is calculated based on the length of the light-clock (L) and the distance the light-clock moves relative to the stationary observer during one clock tick, calculated as velocity times time (vt).

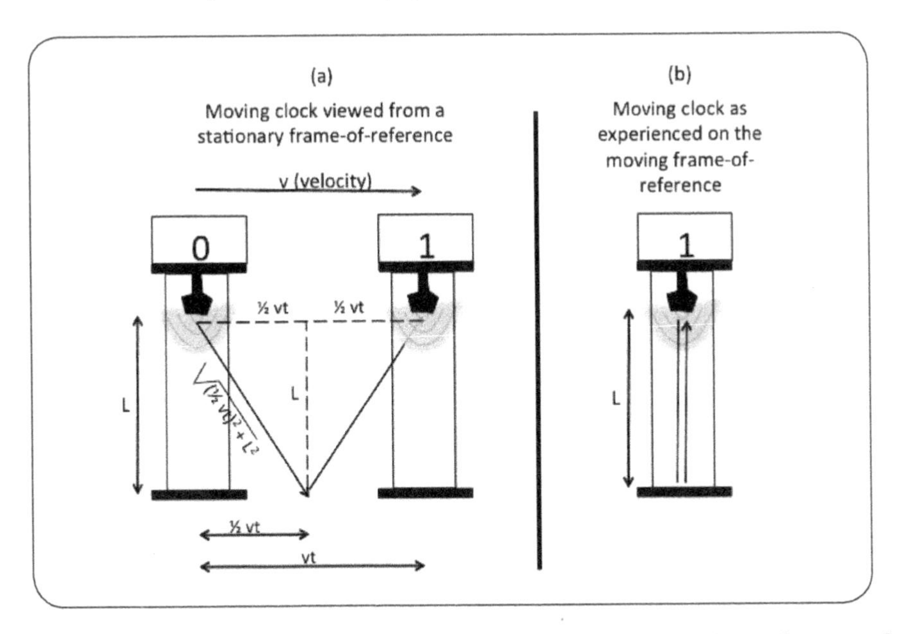

Figure A-2. The Pythagorean theorem is used to calculate the amount of time slowing and space contraction (i.e., the relativistic correction factor). *In this figure, the lengths of the light path on a moving light clock as seen from the stationary frame-of-reference (Side a) and as experienced inside the moving frame-of-reference (Side b) are shown. These can be used to derive the formulas for time slowing and length contraction. The calculations use the formula for distance (distance = velocity x time) and the Pythagorean theorem.*

Now let's use the Pythagorean theorem to calculate the amount of time slowing and space compression that occurs from Special Relativity.

Experienced on the moving frame-of-reference (Side b in the figure):

First, we calculate the time it takes for the light to cover the distance of one clock tick (light goes from top to bottom of clock and back to the top) as experienced inside the moving frame-of-reference; we'll call this $t_{(moving)}$ (Figure A-2b). Assuming the light travels at speed "c" and since distance = velocity x time:

$$2L = ct_{(moving)}$$

So then:

$$t_{(moving)} = \frac{2L}{c}$$

As seen from the stationary frame-of-reference:

Now let's look at the moving clock as seen from the stationary frame-of-reference in Figure A-2a to find the time it takes the light to complete one clock tick; we'll call this $t_{(stationary)}$ since it is the time as seen from the non-moving frame-of-reference. If the moving light clock is moving at velocity "v," the distance the clock appears to travel is $vt_{(stationary)}$. Using half this distance to make a right triangle, we can use the Pythagorean theorem to calculate the distance the light travels during one half of the clock tick (from the light source to the bottom of the light-clock shown in Figure A-2a).

$$\sqrt{\left(\frac{1}{2}vt_{(stationary)}\right)^2 + L^2}$$

Now the total distance the light travels during one complete clock tick is twice this value. And since the light travels at speed "c" during the entire time of the clock tick, the formula becomes:

$$ct_{(stationary)} = 2\sqrt{\left(\frac{1}{2}vt_{(stationary)}\right)^2 + L^2}$$

$$ct_{(stationary)} = 2\sqrt{\frac{1}{4}v^2t_{(stationary)}{}^2 + L^2}$$

Now we find the relationship between $t_{(stationary)}$ and $t_{(moving)}$. This will tell us how much time is slowed by Special Relativity. The following steps do this.

First, square both sides in the above formula (to get rid of the square root sign):

$$c^2t_{(stationary)}{}^2 = 4\left(\frac{1}{4}v^2t_{(stationary)}{}^2 + L^2\right)$$

$$c^2t_{(stationary)}{}^2 = v^2t_{(stationary)}{}^2 + 4L^2$$

Subtract $v^2t_{(stationary)}{}^2$ from both sides:

$$c^2t_{(stationary)}{}^2 - v^2t_{(stationary)}{}^2 = 4L^2$$

Divide both sides by c^2

$$t_{(stationary)}{}^2 - \frac{v^2}{c^2}t_{(stationary)}{}^2 = \frac{4L^2}{c^2}$$

Factor out $t_{(stationary)}{}^2$

$$t_{(stationary)}{}^2\left(1-\frac{v^2}{c^2}\right)=\frac{4L^2}{c^2}$$

Take the square root of both sides.

$$t_{(stationary)}\sqrt{\left(1-\frac{v^2}{c^2}\right)}=\frac{2L}{c}$$

But $\frac{2L}{c}$ is the time as experienced on the moving frame $t_{(moving)}$. Therefore, we have the formula for the time slowing effect of Special Relativity; that is, we have the formula that relates time in the moving frame $t_{(moving)}$ with time in the stationary frame $t_{(stationary)}$.

$$t_{(moving)}=t_{(stationary)}\sqrt{\left(1-\frac{v^2}{c^2}\right)}$$

Finally, we need to figure out the amount of space compression. This is easy since the amount of space compression is directly related to the amount of time slowing. This is because distance is velocity times time ($d = vt$). We just have to multiply both sides of the equation by velocity to get distance. This is shown below.

$$vt_{(moving)} = vt_{(stationary)}\sqrt{1-\frac{v^2}{c^2}}$$

$$d_{(moving)} = d_{(stationary)}\sqrt{1-\frac{v^2}{c^2}}$$

Finally, we can see that to find the amount of time slowing or the amount of space compression in a moving frame-of-reference, we simple need to multiply by the relativistic correction factor:

$$\sqrt{1-\frac{v^2}{c^2}}$$

Appendix B

Calculations for the Addition of Relative Velocities Considering the Effects of Special Relativity

For those interested, this appendix addresses the following questions:

B.1 Adding relativistic velocities. *How do we add velocities together in Einstein's world? How does Special Relativity make it mathematically impossible to exceed the speed of light?*

B.2 Adding everyday velocities. *Why doesn't Special Relativity matter at slow speeds? What happens when we apply Special Relativity corrections for slow speed velocities? What does the math say?*

B.3 The Lorentz transformations. *What are the Lorentz transformations? Why are they important for Special Relativity? How do the formulas for shortened space and slowed time calculated in Appendix A relate to the formulas created using the Lorentz transformations?*

B.4 Deriving the relativity velocity addition formula. *How were the formulas for combining velocities associated with different frames-of-reference derived? How were the Lorentz transformations used to figure out how to combine velocities across different frames-of-reference?*

B.5 Considerations for General Relativity. *How do the transformations for space, time and movement described for Special Relativity relate to General Relativity?*

This appendix presents the mathematical foundations behind Einstein's Special Relativity (at a very simplified and descriptive level). It describes how the main principles behind Special Relativity play out numerically. It expands from the basic formulas derived in Appendix A to describe more completely how velocities can be combined across multiple frames-of-reference. The deeper mathematical basis for these simplified numerical descriptions is also discussed.

B.1 Adding relativistic velocities. This section presents simplified Special Relativity formulas for adding relativistic velocities. These simple formulas are based on the deeper mathematics developed by Einstein and others. Using these formulas we will see why it is not possible to go faster than the speed of light.

As noted throughout this book, Special Relativity is about measuring time and distance across frames-of-reference that are in uniform motion relative to each other. Special Relativity can also be applied to the more complicated task of determining relative ***velocities*** of objects that are associated with different frames-of-reference. As you would expect, these calculations are more complicated than for just figuring out changes in distance and time between two moving frames-of-reference, as we did in Chapter 6 and Appendix A.

The calculations for combining velocities across multiple frames-of-reference can be visualized as simply adding the velocities together after applying the relativistic correction factor to account for the time slowing and space contraction effects of Special Relativity experienced within each frame-of-reference. Figure B-1 shows a hypothetical example of adding high-speed velocities described earlier, to illustrate.

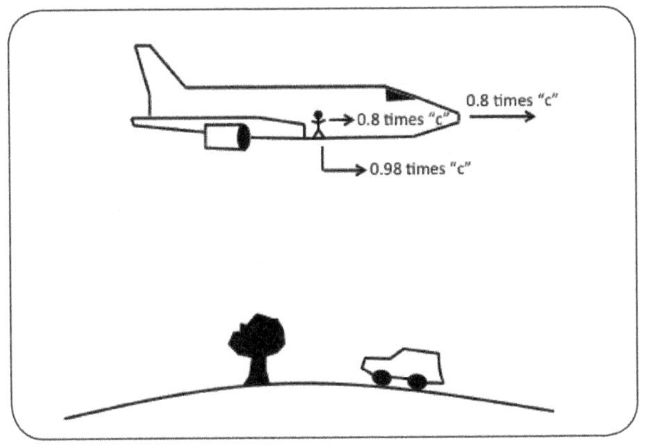

Figure B-1. Adding velocities considering the effects of relativity (nested example). *If an airplane were to fly 80% of the speed of light and one could walk down the aisle at 80% of the speed of light, the combined speed relative to the Earth considering Special Relativity would be 98% of the speed of light.*

So let's look at how we add the velocities of the plane and person. Common sense would hold that the speed of the person relative to the Earth (V_{total}) would equal the sum of the person's speed inside the airplane (V_{person}) plus the speed of the airplane over the Earth ($V_{airplane}$) as shown in the formula below:

$$V_{total} = V_{airplane} + V_{person}$$

But as noted earlier, this doesn't account for the effects of Special Relativity. Recall that the relativistic correction factor that was calculated using the Pythagorean theorem in Appendix A is:

$$\sqrt{1 - \frac{v^2}{c^2}}$$

This means that the calculations for distance and time on moving frames-of-reference are:

$$d_{moving} = d_{stationary}\sqrt{1 - \frac{v^2}{c^2}}$$

$$t_{moving} = t_{stationary}\sqrt{1 - \frac{v^2}{c^2}}$$

But these are simplified formulas for just calculating distance and time within each moving frame-of-reference as seen from the other. They don't work for adding the velocities relative to a third frame-of-reference selected as "stationary". The relativistic correction factors for both time and distance have to be applied before velocities of the moving things (i.e., airplane and person) can be added. That is, velocities of the plane and person have to be adjusted for slowed time and shortened space before they can be combined.

Einstein used the Lorentz transformation equations developed by Dutch physicist Hendrik Lorentz for making these calculations in a way that accounts for the effects of Special Relativity on each object within its own frame-of-reference. When the Lorentz transformation equations are applied, the resulting formula for combining velocities across multiple frames-of-reference is shown below for the above example, with Earth as the "stationary" frame-of-reference. This is called "relativistic velocity addition formula".

$$V_{total} = \frac{V_{airplane}+V_{person}}{1+\frac{(V_{airplane})(V_{person})}{c^2}}$$

For those interested, a description of how this formula is derived using the Lorentz transformation equations is given in Section B.4 of this appendix.

So let's use the relativistic velocity addition formula, given above, and work through some examples of adding the velocities to see how it works, and to see how velocities get restricted to speed "c," mathematically. We will use the examples from Chapter 9.

Example 1. "Nested" relativistic speeds. Here is the "nested" example with speeds approaching speed "c" from Chapter 9. The conditions for this example are reproduced in Figure B-1 (above). The velocities to be added are.

$V_{airplane} = speed\ of\ the\ airplane\ (as\ a\ \%\ of\ "c")$

$= 80\%\ (i.e., 0.80{\times}"c")$

$V_{person} = speed\ of\ person\ on\ the\ airplane\ (as\ a\ \%\ of\ "c") =$

$80\%\ (i.e., 0.80{\times}"c")$

$"c" = speed\ of\ light\ (as\ a\ \%\ of\ "c") = 100\%\ (i.e., 1\ x\ "c")$

The total combined speeds (V_{total}) relative to Earth, the frame-of-reference selected as "stationary," considering the effects of relativity, are:

$$V_{total} = \frac{0.8c + 0.8c}{1 + \frac{(0.8c)(0.8c)}{c^2}} = \frac{1.6c}{1 + \frac{.64c^2}{c^2}} = \frac{1.6c}{1 + .64} = \frac{1.6c}{1.64} = 0.976c$$

So, the combined speed relative to Earth expressed as a proportion of speed "c" is 0.976 x "c" or 97.6% of speed "c."

Example 2. "Approaching" relativistic speeds. Here is the "approaching" example from Chapter 9, and reproduced in Figure B-2. This time the formula is:

$$V_{total} = \frac{V_{airplane\ 1} + V_{airplane\ 2}}{1 + \frac{(V_{airplane\ 1})(V_{airplane\ 2})}{c^2}}$$

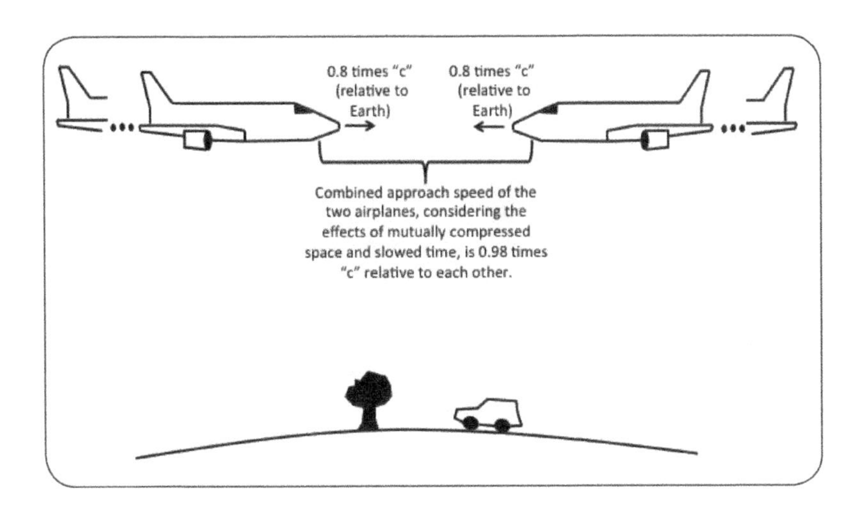

Figure B-2. Adding velocities considering the effects of relativity (approaching example). *If two airplanes were to fly towards each other at 80% of the speed of light, the combined speed relative to Earth considering Special Relativity would be 98% of the speed of light.*

The velocities to be added for this example are:

$V_{airplane\ 1} = speed\ of\ the\ airplane\ (as\ a\ \%\ of\ "c")$

$= 80\%\ (i.e., 0.80 \times "c")$

$V_{airplane\ 2} = speed\ of\ the\ second\ airplane\ (as\ a\ \%\ of\ "c") =$

$80\%\ (i.e., 0.80 \times "c")$

$"c" = speed\ of\ light\ (as\ a\ \%\ of\ "c") = 100\%\ (i.e., 1 \times "c")$

The combined speeds (V_{total}), relative to Earth, the frame-of-reference selected as "stationary," considering the effects of relativity are:

$$V_{total} = \frac{0.8c + 0.8c}{1 + \frac{(0.8c)(0.8c)}{c^2}} = \frac{1.6c}{1 + \frac{0.64c^2}{c^2}} = \frac{1.6c}{1 + 0.64} = \frac{1.6c}{1.64}$$

$$= 0.976c$$

So, the combined speed relative to Earth expressed as a proportion of speed "c" is 0.976 x "c", or 97.6% of speed "c" (same as for the "nested" case in Example 1).

Example 3. Photons approaching other photons. Here is the case for photons approaching other photons. This is illustrated in Figure B-3.

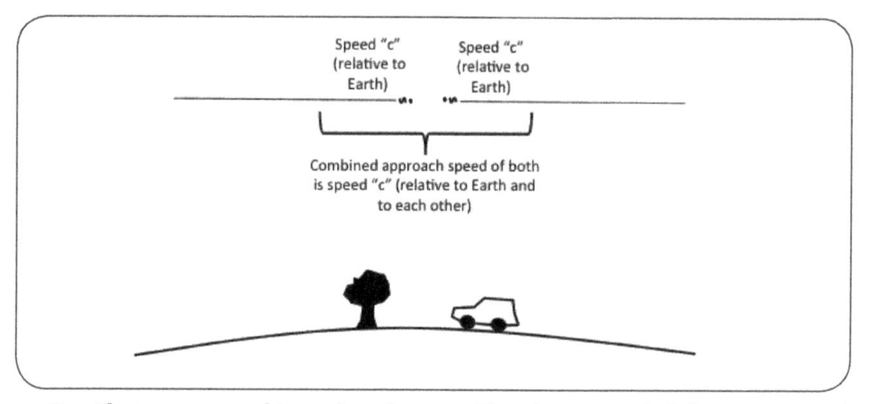

Figure B-3. Photons approaching other photons. *The relative speed of photons approaching other photons.*

$$V_{photon\ 1} = speed\ of\ the\ first\ photon\ (as\ a\ \%\ of\ \text{c})$$

$$= 100\%\ (i.e., 1\times\text{"c"})$$

$$V_{photon\ 2} = speed\ of\ the\ second\ photon\ (as\ a\ \%\ of\ \text{"c"}) =$$

$$100\%\ (i.e., 1\times\text{"c"})$$

$$\text{"c"} = speed\ of\ light\ (as\ a\ \%\ of\ \text{"c"}) = 100\%\ (i.e., 1\times\text{"c"})$$

This time, the formula is:

$$V_{total} = \frac{V_{Photon\ 1} + V_{photon\ 2}}{1 + \frac{(V_{photon\ 1})(V_{photon\ 2})}{c^2}}$$

The combined speeds (V_{total}), relative to Earth, the frame-of-reference selected as "stationary," considering the effects of relativity, are:

$$V_{total} = \frac{1c+1c}{1+\left(\frac{(1c)(1c)}{c^2}\right)} = \frac{2c}{1+\frac{c^2}{c^2}} = \frac{2c}{1+1} = \frac{2c}{2} = c$$

So, the combined speed of two photons approaching each other expressed as a proportion of speed "c," is 1 x "c" or speed "c."

B.2 Adding everyday velocities.

This section uses the relativistic velocity addition formula to demonstrate why the effects of Special Relativity are negligible for our everyday lives.

As noted in this book, Special Relativity applies to all uniform motion. However, the movements we experience on Earth, even on fast airplanes, are so slow compared to the speed of light that the impact of Special Relativity is extremely small. The next example illustrates this mathematically.

Example 4. Normal speeds. Here is the nested example with normal airplane and walking speeds from Chapter 9. This example, reproduced in Figure B-4, shows that the effects of relativity for normal speeds are so small that they are negligible and can be ignored.

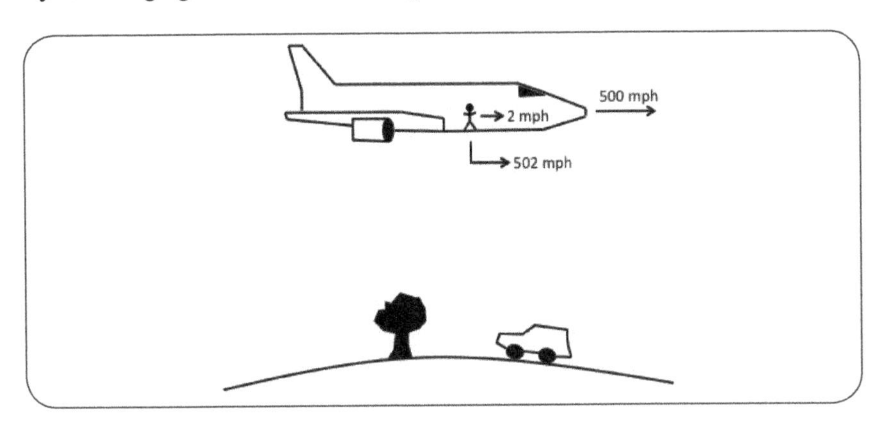

Figure B-4. Adding velocities for normal speeds considering the effects of relativity (nested example). *For an airplane flying 500 mph and a person walking 2 mph, the combined speed relative to the Earth considering Special Relativity would be 502 mph.*

The velocities to be added are:

$$V_{airplane} = speed\ of\ the\ airplane = 500\ mph = 0.0000007\times"c"$$

$$V_{person} = speed\ of\ the\ person\ on\ the\ airplane = 2\ mph$$

$$= 0.000000003\times"c"$$

$$\mathrm{c} = speed\ of\ light\ (as\ a\ \%\ of\ \mathrm{c}) = 100\%\ (i.e., 1\times"c")$$

As before, the formula is:

$$V_{total} = \frac{V_{airplane}+V_{person}}{1+\frac{(V_{airplane})(V_{person})}{c^2}}$$

The combined speed (V_{total}), relative to Earth, the frame-of-reference selected as "stationary," considering the effects of relativity, are:

$$V_{total} = \frac{0.0000007c + 0.000000003c}{1 + \frac{(0.0000007c)(0.000000003c)}{c^2}} = \frac{0.000000703c}{1.000000000000002}$$

$$= 0.000000703c = 502\ mph$$

This example shows that the combined speed, expressed as a proportion of speed "c," for normal airplane and walking speeds is 0.000000703 x "c." Since this is the same as the sum of the airplane speed and walking speed (502 mph), it can be seen that the effects of relativity are so small that they are negligible and can be ignored, which is what we do every day. If we were to travel at speeds closer to the speed of light, we would learn to recognize and account for the effects of relativity.

B.3 The Lorentz transformations. This section presents a description of the Lorentz transformations as applied to Special Relativity and a discussion of why they're important. It also presents the fundamental Lorentz transformation equations for distance and time, and shows how these equations are mathematically related to the relativistic velocity addition formula.

OPTIONAL: For those interested, this section can help you to understand the relationship between the Lorentz Transformation equations.

As noted above, the Lorentz transformations are needed for determining relative velocities of objects that are associated with different frames-of-reference. The way the Lorentz transformations do this is by first defining ***space-time events*** as the locations of objects in (integrated) space-time. This allows evaluating the motion of objects as ***space-time events*** that change their space-time locations relative to space-time, and therefore relative to ***all*** frames-of-reference. The events are defined as coordinates x, y, and z (for the spatial dimensions) and t for the time coordinate. These x, y, z, t space-time event coordinates change as an object moves. Since space (x, y, z) and time (t) are part of the event location definition, the changing space-time locations automatically capture any velocities. When an object moves, it defines a new space-time event with the new coordinates x*, y*, z*, t*. These are really dynamic events with continuously changing values for x, y, z, and t.

The x, y, z, t values naturally reflect any selected frames-of-reference across which their movements are being measured. The trick is to relate these dynamic events with changing values for x, y, z, and t across all the frames-of-reference of interest. This automatically captures the effects of the relative motion between selected frames-of-reference. Since the thing that is common and unchanging across ALL frames-of-reference is the speed of light, speed "c," the Lorentz equations use this constant when defining the new x*, y*, z*, t* coordinates of the different frames-of-reference. The Lorentz transformations use the constant of speed "c" to derive equations that can relate all space-time events across all frames-of-reference. Einstein's introductory book on relativity provides a very interesting description of the concept with appendices that summarize underlying derivation of the Lorentz transformations.[1]

B.4 Deriving the Relativistic Velocity Addition Formula. This section summarizes how the Lorentz Transformation equations form the foundation for the relativistic velocity addition formula.

OPTIONAL: For those interested, this section describes how the Lorentz Transformation equations are applied to develop the relativistic velocity addition formula.

So, how are the Lorentz transformations used to develop the formula for combining velocities across frames-of-reference? This is the formula we used for the velocity combining examples discussed earlier in this appendix. This formula can be used for combining velocities associated with multiple frames-of-reference, including when the velocities are nested like in Example 1 and when velocities are being added like in Examples 2 and 3, as long as the velocities being added are in motion along just one of the spatial dimension, for example the x-dimension. When the velocities to be added involve motion

1 See Chapter 26 in Einstein's introductory book on Relativity called *Relativity: The Special and General Theory*. A more complete mathematical discussion is available in Robert Resnick's book *Introduction to Special Relativity*. The math in this section is based on Resnick's book.

across more than just one spatial dimension, say x and y, more complicated formulas are needed[2]. This simplified formula is used here as in other books on relativity to show the concept.

The following is a summary of how this simplified relativistic velocity addition formula was developed for Example 1, the nested example.

We start with the formulas for: (1) *the* ***time elapsed*** and (2) *the* ***distance traveled*** *in a moving frame-of-reference compared to what is observed from a "stationary" frame-of-reference. These formulas flow from Special Relativity and the mathematics of the Lorentz transformations.*

Relevant variables are:

v = Velocity of the moving frame-of-reference relative to the stationary frame-of-reference.
t_S = Time elapsed on the stationary frame-of-reference.
t_M = Time elapsed on the moving frame-of-reference. (Due to Special Relativity – "*moving clocks run slow.*")
d_S = Distance traveled as observed from the stationary frame-of-reference.
d_M = Distance traveled as experienced on the moving frame-of-reference. (Due to Special Relativity.)
c = Speed of light.

Also, here are the Lorentz Transformation formulas that use these variables. We will apply these equations as we work through our example[3].

Lorentz Transformation Equation for Time:

EQUATION 1

$$t_M = \frac{t_S + \frac{v}{c^2} d_S}{\sqrt{1 - \frac{v^2}{c^2}}}$$

2 See discussion starting on Page 79 of Resnick's book, *Introduction to Special Relativity*, for a description of these more complex calculations, including the addition of velocities that are nested in a moving frame-of-reference, and those moving directly relative to each other relative to a third frame-of-reference, as well as motion involving more than just one spatial dimension.

3 See Appendix A of Einstein's book, *Relativity: the Special and General Theory* to see how the Lorentz Transformation equations are derived.

Lorentz Transformation Equation for Distance:

EQUATION 2

$$d_M = \frac{d_S + vt_S}{\sqrt{1-\frac{v^2}{c^2}}}$$

Notice that both of these formulas contain the relativistic correction factor that we developed using the Pythagorean theorem in Appendix A:

$$\sqrt{1-\frac{v^2}{c^2}}$$

But now we are considering velocities that occur within one of the respective frames-of-reference . We therefore need to consider the Special Relativity effects of movement within the moving frame-of-reference, as well as the Special Relativity effects between the moving frame-of-reference and the stationary frame-of-reference.

Let's begin by specifying an object that is moving within the moving frame-of-reference. Let's say the object is traveling at velocity **w** within a moving frame-of-reference, then since ***distance*** = ***velocity*** x ***time***, we get:

EQUATION 3

$$d_M = wt_M$$

Now, if we use the Lorentz transformation, Equation 1 to substitute for t_M and Lorentz transformation, Equation 2 to substitute for d_M in Equation 3, we get:

EQUATION 4

$$\frac{d_S + vt_S}{\sqrt{1-\frac{v^2}{c^2}}} = w\left(\frac{t_S + \frac{v}{c^2}d_s}{\sqrt{1-\frac{v^2}{c^2}}}\right)$$

Since Equations 1 and 2 account for the relativity effects of the moving frame-of-reference, it makes sense to apply them here, so velocity **w** will reflect the Special Relativity effects of the moving frame-of-reference as observed from the "stationary" frame-of-reference.

Now to solve for **w**, we multiply both sides of Equation 4 by the reciprocal of the expression in parentheses. This simplifies to equation 5:

EQUATION 5

$$\frac{d_S+vt_S}{t_S+\frac{v}{c^2}d_S}=w$$

We can now use Equation 5 to derive an equation for combined velocity, V, of an object that is moving inside a moving frame-of-reference.

And since, the object moving within the moving frame-of-reference, as seen from the stationary frame-of-reference, is described by the formula $d_s = wt_s$, we can make this substitution for d_s and arrive at the formula for combined velocity, V:

$$V=\frac{wt_S+vt_S}{t_S+\frac{v}{c^2}wt_S}$$

Factoring and reducing:

$$V=\frac{t_S(w+v)}{t_S(1+\frac{v}{c^2}w)}$$

$$V=\frac{w+v}{1+\frac{v}{c^2}w}$$

EQUATION 6

$$V=\frac{w+v}{1+\frac{vw}{c^2}}$$

So substituting in Example 1, we get:

$$V_{total}=\frac{V_{airplane}+V_{person}}{1+\frac{(V_{airplane})(V_{person})}{c^2}}$$

B.5 Considerations for General Relativity. This section introduces the issues involved in extending the above equations and considerations to the broader General Theory of Relativity.

The above formulas and calculations are for Special Relativity. They therefore assume that objects and frames-of-reference all move uniformly and see space-time as rigid and unbending. But we know that space-time is really bent by gravity, which permeates the whole Universe, and that objects experience accelerations. Special Relativity provides a good approximation for the equations of General Relativity when the gravitational fields are weak and accelerations are absent. It therefore represents a "special case" within General Relativity defined by weak gravity and the absence of accelerations.

While the simplifying assumptions of Special Relativity make it easier to see how things work, and the math more understandable, they do not consider all the complexities of the real Universe. Space-time is a flexible thing that bends near gravity. This bending of space-time affects both space and time, making the consideration of space and time using a rigid (i.e., unbending space-time) framework inaccurate, especially where gravity is strong. Non-rigid frames-of-reference (e.g., bent by gravity) are common throughout the Universe and are described by the more complex mathematics of General Relativity. Einstein called these flexible (non-rigid) reference frames "reference mollusks." Einstein said that it must be possible to define any reference mollusk as the selected "stationary" frame-of-reference. This can only be accomplished with the more difficult formulations of General Relativity. The ability to handle the complex dynamics of objects that move through space-time that is bent by gravity and accelerations illustrates the power that General Relativity brings to the science of physics.

It is these complexities of how space-time works in the real Universe that creates the difficulty in accomplishing General Relativity mathematically. This is why General Relativity required a new kind of geometry and complicated tensor calculus to make it work. An introduction to General Relativity is provided in Einstein's book "Relativity: The Special and General Theory."

Appendix C
How to Remember Relativity Principles

This appendix addresses the following questions:

C.1 Reviewing the pieces together. *How can we put the concepts of the Special and General Theories of Relativity into an integrated context for better remembering? How can we better see the big picture and gain a more intuitive sense of how relativity works?*

C.2 Exercises for remembering. *How can we use everyday experiences to help internalize our understanding of relativity? Are there routine exercises that can help us more deeply understand and appreciate the concepts of relativity?*

If you find it difficult to fully grasp and relate to relativity on a personal and intuitive level, join the club. Even scientists who teach relativity express difficulty in being fully able to appreciate it in a common sense and experience-based way. This is why they use mnemonics like *"moving clocks run slow"* to help remember how it works. This is understandable since the conditions that we experience in our everyday lives do not come close to the conditions in which the effects of relativity become significant and would be noticeable. We therefore don't experience the effects of relativity in our everyday lives. In fact, we never experience them, ever.

In spite of this shortcoming, it is possible to superimpose imagined conditions on our everyday experiences to create situations where the relativity effects would be significant and can be envisioned. In this way, it is possible to build a conceptual understanding of how relativity works without actually experiencing relativity-producing conditions. This appendix provides simple exercises to help develop a deeper understanding of relativity that will be easier to remember.

But first, it is helpful to step back and review the relativity big picture; to see, in a single frame, why relativity works. This integrated summary review will help internalize the relativity concepts to gain a more intuitive sense of what's involved and why relativity works.

C.1 Reviewing the pieces together. Reviewing how the elements of relativity fit together, and the scientific discoveries that make relativity the best explanation for why the Universe works the way it does, can strengthen our understanding of the relativity principles. This section presents such an integrated summary by bringing together the main points from the preceding chapters.

A good place to start in reviewing the principles of relativity is in Table 1-1 in Chapter 1 of this book. Table 1-1 introduces the concepts covered in this book. It also serves as a good resource for reviewing the concepts that were covered. The table covers the contents of this book and presents the interrelated discoveries, and their contexts, that define relativity on just a few pages. It therefore provides a way for quickly reviewing the concepts.

Review of the Special Theory of Relativity. You will recall that three key observations about how the Universe works led Einstein to propose his Special Theory of Relativity. These are:

1. All motion is relative.
2. There is no universal frame-of-reference in space against which to measure motion.
3. The speed of light is always the same regardless of any relative motion between light source and observer.

When these facts are considered in combination, they have unavoidable ramifications for time and space, and they combine to form the foundation for Special Relativity. Specifically, when we measure the speed of light, we find that it remains the same, even when the light source is moving. This fact together with the knowledge that velocity is simply distance per time (e.g., miles per hour) led Einstein to the conclusion that it must be time that is changing! That is, it must be time itself that moves more slowly on moving platforms, and space must be correspondingly compressed (a shortened light path) for objects in relative motion. This is the point made in Chapters 2, 3, 4, 5 and 6. It is summarized in Figure C-1. If you aren't recalling the concepts shown in the figure, consider going back and reviewing the illustrations in the appropriate chapters.

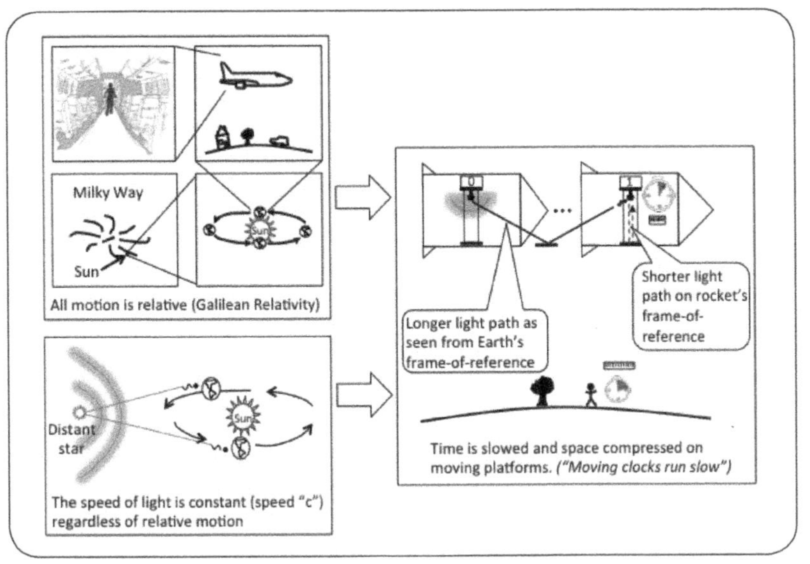

Figure C-1. Integrated view of Special Relativity. *The facts that: (1) all motion is relative; (2) there is no universal frame-of-reference in space; and (3) the speed of light is always the same regardless of any relative motion of the light source, all form the foundation for Special Relativity. It means that time is slower and space is compressed on platforms that are in uniform (non-accelerating) relative motion.*

But Special Relativity has limitations. The simple relationships that hold under Special Relativity fall short when conditions of acceleration and gravity are present. Acceleration, like that created from an external force, such as a jet engine; or when fighting against the pull of gravity, like when standing on the surface of the Earth, make Special Relativity inadequate for fully describing the relationship between time and space. This limitation led Einstein to develop the more comprehensive General Theory of Relativity, summarized below.

Review of the General Theory of Relativity. Situations involving acceleration are described by Einstein's more complete and comprehensive solution, called the Theory of General Relativity. The distinctions between accelerating and non-accelerating conditions, and the respective impacts on light, space and time as described by Special and General Relativity are summarized in Figure C-2. More complete descriptions of these effects are provided in Chapters 4 and 8. The ramifications of both Special and General Relativity for the Universe as a whole are summarized in Chapters 9 and 10.

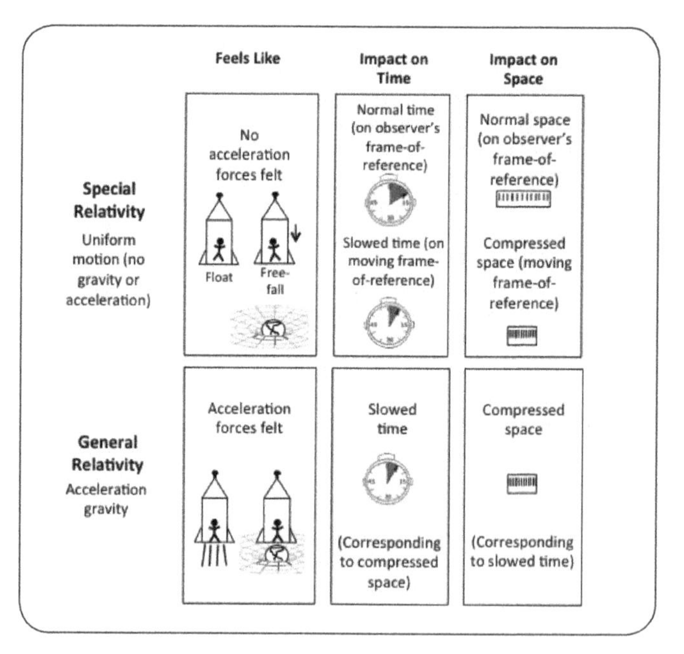

Figure C-2. Integrated view of Special and General Relativity. *Under conditions of acceleration, the solutions of Special Relativity fall short. The more complete solution offered by the General Theory of Relativity is needed. General Relativity describes how acceleration and gravity affect space and time.*

Figure C-2 summarizes some of the key distinctions between Special and General Relativity. Like all illustrations in this book, these images are simplified two-dimensional depictions to help us visualize the conditions and effects. The underlying reality is much more complicated, described with advanced mathematics and a different way of representing geometry. The two-dimensional pictures used in this book help us visualize the concepts of relativity within the three-dimensional geometry we are familiar with.

C.2 Exercises that can turn everyday experience into relativity classrooms. To internalize and understand the concepts of relativity, it helps to experience them regularly. This section describes how everyday life situations can be enhanced with imagined conditions to provide regular repeated experience with the effects of relativity.

Our everyday movements and experiences have all the ingredients necessary for seeing the effects of relativity. They just don't occur under sufficiently extreme conditions for the relativity effects to be noticed. But it is easy to apply imagined conditions to everyday events that allow the effects of relativity to be experienced in a virtual way. This chapter describes exercises that you can do every day to transform normal events into hands-on classrooms for better understanding relativity. These exercises apply to both Special Relativity and General Relativity. They can be done while simply going about our everyday lives. They involve superimposing imagined conditions onto the everyday events, making the effects of relativity more obvious. This will not only help appreciate how relativity works but, will build a deeper understanding of relativity's principles.

First, we'll examine how everyday experiences can be enhanced with imagined conditions in a way that builds a more intuitive understanding of Special Relativity. The biggest impediment to experiencing the effects of Special Relativity during our normal movements is the slow relative speeds involved (compared to the speed of light). The speeds of even our fastest airplanes are far too slow for the effects of Special Relativity to be evident. Since we are not capable of experiencing speeds approaching the speed of light, not to mention the danger involved in moving at such high speeds, it is necessary to use our imaginations to visualize these extreme conditions and the resulting Special Relativity effects. Using this approach, we can experience the effects of Special Relativity every day. We can do this by mentally exaggerating the motions we do experience, thus superimposing imaginary conditions under which the effects of Special Relativity are obvious or at least imaginable.

When doing these exercises, remember the basic conditions that underscore the principles of relativity: light always travels at speed "c" regardless of relative motion of the light source; speed equals distance traveled divided by the time elapsed. This means that with a constant, unvarying speed of light, it is time and space themselves that change when objects move relative to each other.

The first exercise involves watching an airplane fly across the sky and imagining conditions that would make the effects of Special Relativity discernable. Imagine that the airplane you see is moving at speeds near the speed of light; say 75% of speed "c." Visualize a light-clock on the airplane to see how a photon would appear to travel further from your Earth-bound frame-of-reference compared to the same photon as experienced by someone on the plane. With the light's speed constant (speed "c") in both frames-of-reference, it is time and space on the airplane that is altered. The shorter light path experienced on the moving airplane equates to compressed space and slowed time as seen from Earth. You may have to do this exercise many times before you get an intuitive sense of how Special Relativity works. This is shown in Figure C-3.

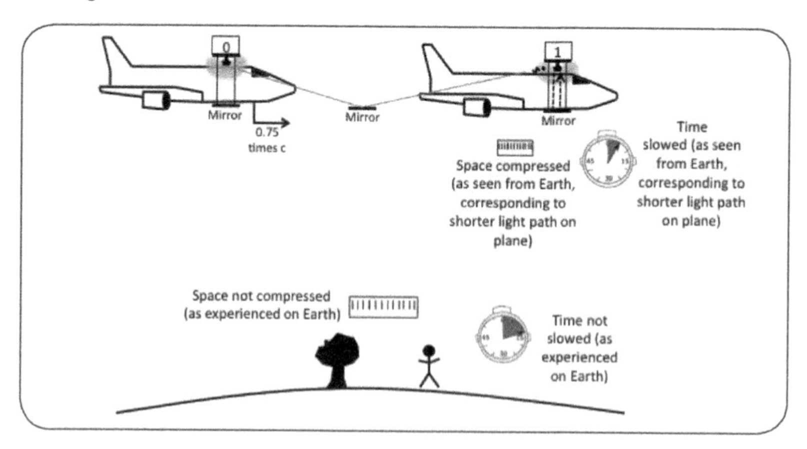

Figure C-3. Imagining the effects of Special Relativity (high speed travel). *Imagine that a passing airplane is traveling at speeds approaching the speed of light, say 75% of speed "c." Visualize a light-clock onboard the plane to see how time is slowed, corresponding to a compressed space on the plane's frame-of-reference.*

Now to get a sense of how the amount of time slowing and space compression is affected by the amount of relative speed involved, imagine that the plane is only moving at 50% of the speed of light. Again imagine a light-clock onboard. This time (at the lower speed) the effects of Special Relativity are less. There is less time slowing and less space compression. This is shown in Figure C-4. While there is a tiny amount of time slowing and space compression associated with the plane's actual speed, say 500 mph, these differences are so small they are insignificant. It is only at speeds approaching the speed of light that time slowing and space compression become significant.

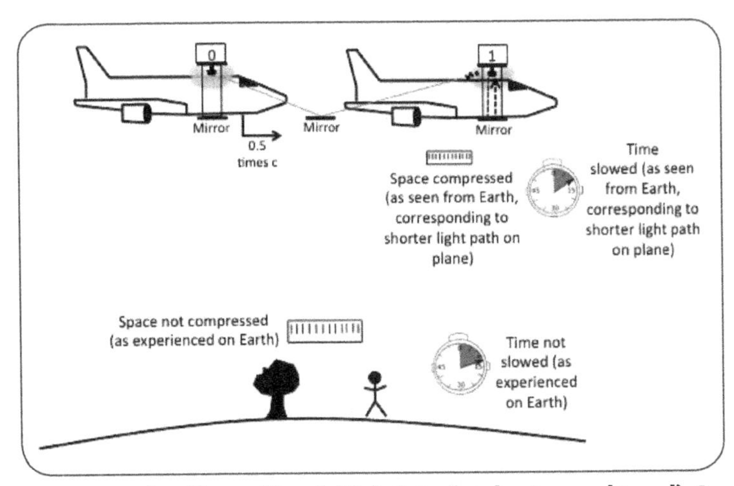

Figure C-4. Imagining the effects of Special Relativity (moderate speed travel). *Imagine that a passing airplane is traveling at a slower speed than in the previous example, but still at a pretty high speed relative to the speed of light; say this time at 50% of speed "c." Again, visualize a light-clock onboard the plane to see how time is slowed corresponding to a compressed space on the plane's frame-of-reference, but this time less compressed space and less slowed time is experienced.*

Now imagine a helicopter hovering overhead. In this case, since there is no motion relative to you on the Earth's surface, there is no time slowing or space compression. This is shown in Figure C-5. Do this exercise when watching moving airplanes and stationary helicopters to become familiar with the effect of relative motion on time and space; that is, to visualize how relative motion has the effect of slowing time and compressing space.

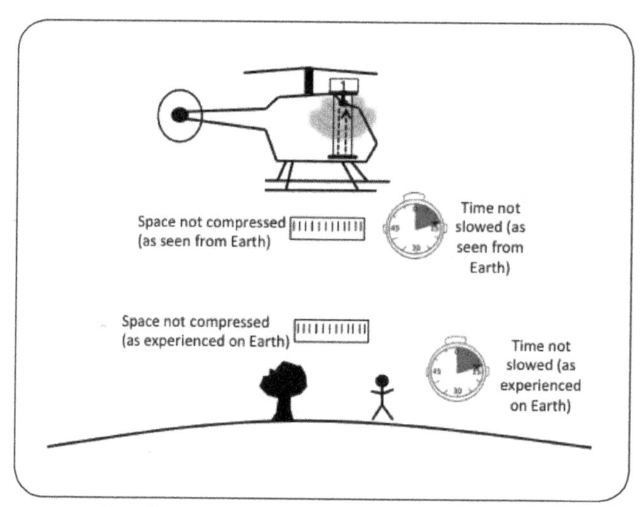

Figure C-5. Imagining the effects of Special Relativity (without relative motion). *Imagine that a helicopter is hovering overhead so there is no relative motion. Visualize a light-clock onboard the helicopter and realize that there is no time slowing or compression of space. You and the helicopter are in the same frame-of-reference.*

These are simple exercises that can be done frequently as a way of becoming familiar with the effects of Special Relativity. They provide a way to visualize the Special Relativity effects, even though it is not possible to directly experience it in real life.

Second, let's look at how some everyday experiences can be used to remind us of how General Relativity affects space and time by bending or compressing the integrated space-time. Like for Special Relativity, it is possible to imagine the effects of General Relativity during normal Earth-bound activities, as a way of becoming familiar with them. Even though Earth's gravitational field is far too weak to bend space-time to an extent that we could notice as changes to time or space, mental exercises can help us imagine the effects. In this way we can experience the effects of General Relativity in a virtual manner. The exercises described below can help us imagine the effects of General Relativity.

You will recall that gravity has the effect of bending or compressing space-time in a way that slows time with a commensurate compressing of space, when two are considered separately. When you move further from the Earth, like going to top of a mountain or tall building, space-time is less compressed than when on Earth's surface. This means that nearest to Earth time moves more slowly. Conversely, further from Earth's surface, time goes faster.

The exercise depicted in Figure C-6 shows how we can imagine the effects of General Relativity when in a tall building. Clocks at the bottom of the building will run more slowly than clocks at the top of the building due to the compressing of space-time by Earth's gravity. Because time moves more quickly at the top of the building you will age more quickly when there. And you will age less quickly when on the lower levels of the building. Similar to the experience of Einstein's travelling twins (due to Special and General Relativity), if you were to live your life at the top of a tall building while your twin lived at the bottom of the building, you would grow old faster living at the top (due to General Relativity). Even though the real impact of Earth's gravity on time and space is negligible and insignificant between the top and bottom of even tall buildings, it is possible to imagine the effects, as a way of remembering how they work. The next time you are at the top of a tall building — or even when you go upstairs in your house or apartment — pretend that you can feel time moving faster accelerating your pace of aging. You may want to spend more time downstairs.

While the effect on time and space in buildings is negligible and insignificant, the effect has been measured with very accurate clocks when taken to the tops of mountains. Nevertheless, it is possible to imagine the effect as a way of practicing and remembering.

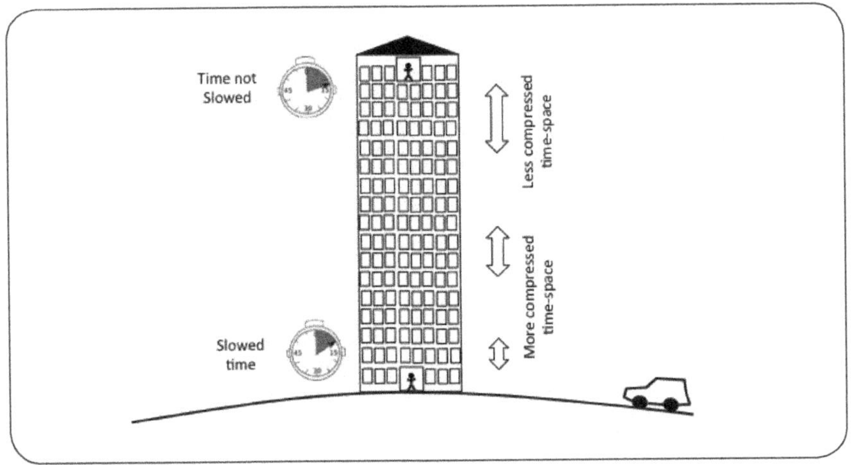

Figure C-6. Exaggerated effect of General Relativity when in a tall building. *Realizing that time is slowed and space compressed near sources of gravity, it is possible to imagine the effect when in tall building. Your watch will run a little faster when at the top of the building compared to when at the bottom and you will age more quickly.*

Also, remember that acceleration has the same effect on time and space as gravity. The two are equivalent, even though there are some differences in how we experience them. Additionally, the two effects can be combined as Einstein did when he rode his bicycle in a curved direction. This means that when you experience acceleration, time is being slowed. Since the effects of acceleration and gravity can be combined the experience of being in an accelerating elevator (up or down) will further slow time. This is shown in Figure C-7. The next time you are accelerating in an elevator remember that space-time is being compressed and time is being slowed. You will age more slowly when experiencing the accelerations from an elevator. Again, the effects are negligible and insignificant for the small accelerations produced by an elevator. However, the elevator experience offers a virtual classroom for imagining and remembering how General Relativity affects time.

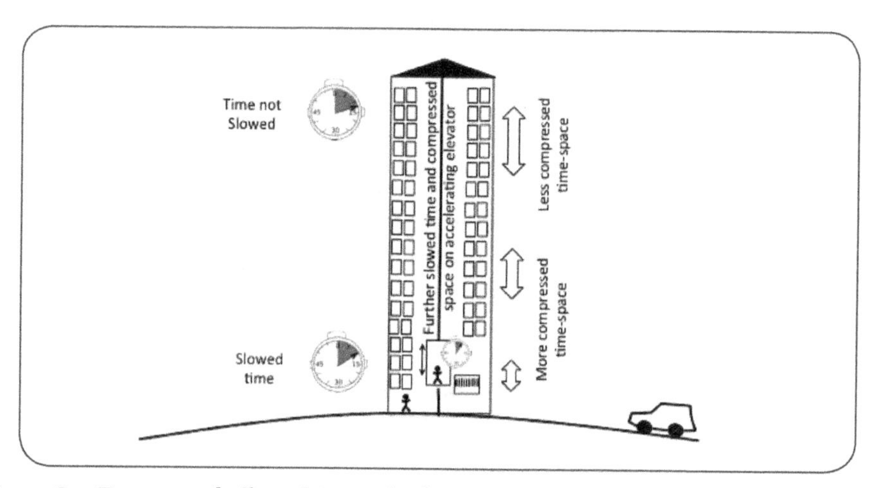

Figure C-7. Exaggerated effect of General Relativity when in an accelerating elevator. *Realizing that time is slowed and space compressed when accelerating, it is possible to imagine the effect of time slowing and space compression when on an accelerating elevator. Your watch will run a little slower and you will age more slowly while the elevator is accelerating (up or down).*

Stop to feel the force of Earth's gravity. Remember that this force is equivalent and indistinguishable from the force of acceleration and that the two can be combined. The force of gravity that you feel on Earth's surface is because your natural, non-accelerating movement through space-time (toward the center of the Earth, in this case) is interrupted by Earth's surface. The energy you expend standing on the Earth's surface is equivalent to the energy you would expend standing on an accelerating rocket ship that creates the same amount of accelerating force. In fact, you can combine the force of gravity with acceleration like Einstein did. When your car makes a sharp turn the force to the side of the car is combining with gravity to temporarily change the direction of down.

One last point to remember as you enjoy Earth's gravity is that the presence of acceleration forces determines whether the laws of Special Relativity are applicable. Uniform motion is a necessary condition for applying the laws of Special Relativity. Acceleration forces from a rocket ship or created by a gravitational field require the more complex laws of General Relativity.

Finally remember that the figures in this appendix, like the figures throughout this book, do not accurately depict how relativity really works. They are drawn on a two-dimensional sheet of paper and cannot fully capture the complex ways relativity operates in four-dimensional space-time. However, they help us visualize, and therefore appreciate, the effects of relativity. They help us understand relativity within the framework of our limited human perspectives. They provide a way to visualize the much more complicated effects of relativity that work in a four-dimensional space-time and use an entirely different mathematical geometry.

References and Sources for Further Reading

Abbott, B.P. et al . "Observation of Gravitational Waves from a Binary Black Hole Merger." *Physical Review Letters*. 116, 061102. 11 February 2016. (http://journals.aps.org/prl/abstract/10.1103/PhysRevLett.116.061102)

Abrams, Nancy Ellen and Primack, Joel R. *The New Universe and the Human Future: How a Shared Cosmology Could Transform the World*. New Haven, CN: Yale University Press. 2011.

Chu, Jennifer. "For Second Time, LIGO Detects Gravitational Waves. Signal was produced by two black holes colliding 1.4 billion years away." *MIT News*. 15 June 2016. (http://news.mit.edu/2016/second-time-ligo-detects-gravitational-waves-0615)

Crelinsten, Jeffrey. *Einstein's Jury: The Race to Test Relativity*. Princeton, NJ: Princeton University Press. 2006.

Einstein, Albert. "Science and Religion." In Einstein, Albert. *Ideas and Opinions*. New York, NY: Three Rivers Press. 1954.

Einstein, Albert. *Relativity: The Special and General Theory*. East Bridgewater, MA: World Publications Group. Revised 1916, reprinted 2007.

Feynman, Richard. *QED: The Strange Theory of Light and Matter*. Princeton, NJ: Princeton University Press. 1985.

Frisch, D. H. and Smith, J. H. "Measurement of the Relativistic Time Dilation Using μ-Mesons." *American Journal of Physics*. 31(5): 342-355. 1963.

Gardiner, Martin. *Relativity Simply Explained*. Mineola, NY: Dover Publications. 1997.

Greene, Brian. *The Elegant Universe: Superstrings, Hidden Dimensions and the Quest for the Ultimate Theory*. New York, NY: Random House. 2000.

Greene, Brian. *The Fabric of the Cosmos: Space, Time, and the Texture of Reality*. New York, NY: Alfred A. Knopf. 2004.

Hafele, J. C. and Keating R. E. "Around the World Atomic Clocks: Predicted Relativistic Time Gains." *Science*, 177, pp. 166-168. July 1972.

Hafele, J. C. and Keating, R. E. "Around the World Atomic Clocks: Observed Relativistic Time Gains." *Science*, 177, pp. 168-170. July 1972.

Hawking, Stephen. *A Briefer History of Time*. New York, NY: Bantam Books. 1999.

Hawking, Stephen and Mlodinow, Leonard. *A Brief History of Time*. New York, NY: Bantam Books. 1998.

Hawking, Stephen and Mlodinow, Leonard. *The Grand Design*. New York, NY: Bantam Books. 2010.

Lisi, A. Garret and Weatherall, James Owen. "A Geometric Theory of Everything." *Scientific American*, December 2010.

NASA. "Ringside Seat to the Universe's First Split Second in the Creation of the Universe." 2016. (nasasearch.nasa.gov/search?utf8=?&affiliate=nasa&query=creation+of+the+Universe)

NASA. "Brief History of the Universe." 2016. (www.nasa.gov/mission_pages/spitzer/multimedia/timeline-2006121889912.html)

Pope Francis. *Laudato Si' On the Care of Our Common Home Encyclical Letter.* 2015.

Pope Francis. *The Light of Faith. Lumen Fidei, Encyclical Letter.* 2013.

Pound, R. V. and Snider, J. L. Effect of Gravity on Nuclear Resonance. *Physical Review Letters.* 13(18): 539-540. 1964.

Primack, Joel R. and Abrams, Nancy Ellen. *The View from the Center of the Universe: Our Extraordinary Place in the Cosmos.* New York, NY: Riverhead Books. 2006.

Resnick, Robert. *Introduction to Special Relativity.* New York, NY: John Riley & Sons, Inc. 1970.

Smolin, Lee. *The Trouble with Physics: the Rise of String Theory, the Fall of a Science, and What Comes Next.* Boston. MA: Houghton Mifflin Harcourt. 2006.

Taylor, Edwin F. and Wheeler, John Archibald. *Spacetime Physics,* 2nd ed. New York, NY: W. H. Freeman. 1992.

Thorne, Kip. *Black Holes and Time Warps: Einstein's Outrageous Legacy.* New York, NY: W. W. Norton. 1994.

Will, Clifford. *Was Einstein Right? Putting General Relativity to the Test.* New York, NY: Basic Books, a Division of HarperCollins. 1986.

Wolfson, Richard. *Simply Einstein: Relativity Demystified.* New York: W. W. Norton. 2003.

A

accelerating frames-of-reference 89, 90

accelerating motion 65, 89, 90, 98

acceleration 2, 7, 41, 62-74, 76-78, 85, 89, 90, 91-99, 101, 103-105, 109, 110-112, 114, 148, 150, 164, 176, 220, 226, 227

aether 6, 23, 24-28, 31, 32, 153, 162, 164

Andromeda Galaxy 17, 121, 138

B

Big Bang 142, 184, 187, 190

black holes 7, 96, 97, 108, 141, 147, 158, 173

Bohm, David 147

C

centrifugal accelerations 66, 67, 69

centrifugal force 7, 63–69

clocks 6, 39–52, 57, 58, 100, 101, 109, 112-114, 119, 131, 150, 162, 168,-170, 173, 174, 218, 225

Copernicus, Nicolaus 65

cosmological constant 177

Creation 9, 175, 180-183, 187, 188, 191-194

D

dark energy 102, 154, 155

dark matter 102, 154

DNA 9, 175, 183, 185-189, 193

DNA (deoxyribonucleic acid) 185

Doppler, Christian 102

Doppler effect 102

"double slit" experiment 144

E

Eddington, Sir Arthur 166, 167

Einstein rings 167

Einstein's relativistic correction factor 55, 56

Einstein's twin paradox 8, 110, 111

electromagnetic frequency shift 170

electromagnetic radiation 5, 13-15, 19, 21, 78, 101, 120, 136, 151, 153, 171

electromagnetic spectrum 14, 22, 101, 102, 106

electromagnetic waves 38, 39, 151, 155, 188

E=mc2 8, 98, 117, 130, 155, 156, 157, 158

entangled particles 178

entanglement 141, 144, 146, 147, 148

Equivalence Principle 7, 63, 69, 71, 89, 164

"Euclidian" mathematics 80

F

Feynman, Richard 14, 139

G

Galilean Relativity 6, 26, 32, 35-36, 64, 66

Galileo 6, 26, 31, 72

General Relativity 2, 5, 7, 10, 32, 65, 75-82, 90, 94, 95, 99, 106, 110, 111, 114, 118, 147, 155, 160, 162, 164-166, 168, 169, 171, 173, 174, 177, 179, 195, 203, 215, 220, 221, 224, 225, 226, 227

General Theory of Relativity 2, 26, 41, 64, 66, 67, 74, 77, 78, 79, 85, 152, 164, 166, 168, 173, 177, 215, 220

Global Positioning System 173

God 9, 175, 179, 181, 183-184, 187, 190, 192, 193

GPS 173–174

Grand Unified Theory 178

gravitational field 69, 89, 92, 94-96, 98, 102, 105, 151-153, 224, 227

gravitational fields 89, 93, 94, 96, 97, 107, 118, 188, 215

gravitational mass 7, 63, 69-72

gravity 2, 7,-9, 33, 62- 78, 85-93, 97, 99, 101, 104-110, 112, 114, 150, 152-155, 158, 160, 162, 164, 169, 171-173, 176, 177, 184, 215, 220, 224, 225-227

gravity waves 9, 152, 153, 155, 171, 172, 173

g-s 69

H

Hawking Radiation 108

Hawking, Stephen 108

Hubble, Edwin 141, 177

I

inertial frame-of-reference 90-92, 94

inertial mass 7, 63, 70-72

inertial motion 41, 89-92

K

kinetic energy 156

L

Laser Interferometer Gravitational-Wave Observatory 171

light 2-8, 10, 13-27, 30, 34-43, 45, 47-51, 54-61, 76, 78, 79, 83-88, 96-99, 101-112, 114-117, 119-122, 125, 126, 129, 132-134, 136- -141, 144, 147-155, 157, 162-171, 173, 176-178, 180, 181, 184, 193, 198-200, 203-205, 207, 209-211, 219-224

light-clocks 48-50, 119, 198-200, 222-224

light-hours 17

light-minutes 17, 137

light waves 14, 15, 19, 20, 25, 101-104, 151, 155, 162, 171

light-year 17, 55, 59, 60, 85, 132, 133, 136

light-years 17, 55, 57, 59-60, 85, 110, 112, 114, 132, 138, 147, 181, 193

LIGO 171-173

Lorentz, Hendrik 78, 206

Lorenz transformations 10, 203, 206, 210-211

M

magnetic fields 151, 152

mass 7, 8, 14, 15, 63, 69-72, 88, 89, 98, 108, 117, 130, 151, 155-158, 171

Maxwell, James Clerk 15, 78

Michelson-Morley experiment 15

Milky Way 16, 17, 30, 108, 181, 193

Minkowski, Hermann 163

Morley, Edward 15

motion 6, 7, 26, 29, 76, 101, 106

Mount Washington 170

moving frame-of-reference 39, 40, 199, 200, 201, 206, 212-214

muons 169-170

P

photons 14-16, 21, 33-34, 39, 42-43, 48, 103, 117-121, 140, 145, 146, 150-152, 154, 155, 169, 184, 208

Photons 14, 16, 117, 140, 159, 208

Poincare, Henri 78

Pope Francis 179

probability waves 146

proper acceleration 96, 98

proper accelerations 94

Pythagorean theorem 10, 43, 55, 60, 198-200, 205, 213

Q

quantum particles 159, 169

Quantum Theory 158-160, 177, 178, 179

R

red shift 104, 167, 168

relative motion 7, 110, 123, 124, 127, 128

relativistic correction factor 10, 54-57, 199, 201, 204-205, 213

religion 9, 175-176, 179-181, 191

rest mass 15, 155

Riemann, Bernhard 80, 99

Riemannian differential geometry 80, 106

Riemannian geometry 80-82, 99

Riemannian mathematics 80

Romer, Ole 15

S

science 9, 158, 160, 169, 175, 176, 179, 180

science and religion 175, 179

shifting of Mercury's orbit 166

simultaneous events 8, 109, 121, 122

Space 6-8, 23, 40, 49, 53, 59, 60, 75, 82, 85, 88, 106, 109, 126, 133-137, 171, 178, 215

Space compression 7, 53, 59

space-time 3, 7-8, 17-18, 26, 69, 75, 76, 80-92, 94-95, 97-99, 101, 104, 107-108, 118-119, 129-138, 141, 143, 147-158, 162-163, 166-167, 170-174, 181, 184, 210-211, 215, 224-227

space-time chart 131-132, 136

space-time compression 171

space-time contours 87, 90, 97, 98

Special Relativity 2, 6-7, 10, 32, 35-36, 41, 47-51, 54, 63-66, 74-75, 77-79, 85, 90, 98, 100, 111, 112, 114-115, 118, 121, 126-128, 156, 163, 170, 174, 198-207, 209-215, 219, 164, 220-224, 227

Special Theory of Relativity 2, 6-7, 10, 26-27, 32, 34-36, 41, 54, 64, 66, 74, 78-79, 85, 116, 156, 219

speed of light 2-3, 8, 10, 13, 15, 17, 18, 19, 24, 25-26, 30, 34, 36-40, 42, 47, 49, 50, 54-61, 78, 84-85, 101, 105, 109-111, 115-117, 119, 125, 126, 133-136, 141, 148-149, 151-152, 157, 163-164, 169-170, 176, 184, 198, 203-205, 207, 209-211, 219, 221-223

String Theory 159-160

T

"tensors" 80

Theory of Invariance 26

Thorne, Kip 97, 147

tidal forces 95-96

time 7, 8, 46, 53-55, 57-58, 60, 61, 97, 106, 109-111, 119, 128, 129, 132, 147, 149, 153, 162, 169, 173, 178, 212

time and aging 8, 109, 110

time dilation 120

time slowing 7, 10, 35, 43, 49-50, 54-55, 58, 60-62, 76, 101, 107, 109, 112, 114, 116, 120, 124-126, 134, 140, 149, 150, 169, 198-199, 201, 204, 223-224, 226

U

uniform motion 34, 41, 65, 74, 79, 85, 89-91, 94, 204, 209

universal speed limit 8, 109, 115, 129, 131, 140

W

wave-particle-duality 14

Wheeler, John Archibald 88

worldline 133-135

wormholes 97, 141, 147, 148